Bonds and Bands
in Semiconductors

MATERIALS SCIENCE AND TECHNOLOGY

EDITORS

ALLEN M. ALPER

GTE Sylvania Inc.
Precision Materials Group
Chemical & Metallurgical
Division
Towanda, Pennsylvania

JOHN L. MARGRAVE

Department of Chemistry
Rice University
Houston, Texas

A. S. NOWICK

Henry Krumb School
of Mines
Columbia University
New York, New York

A. S. Nowick and B. S. Berry, ANELASTIC RELAXATION IN CRYSTALLINE SOLIDS, 1972

E. A. Nesbitt and J. H. Wernick, RARE EARTH PERMANENT MAGNETS, 1973

W. E. Wallace, RARE EARTH INTERMETALLICS, 1973

J. C. Phillips, BONDS AND BANDS IN SEMICONDUCTORS, 1973

In preparation

J. H. Richardson and R. V. Peterson (editors), SYSTEMATIC MATERIALS ANALYSIS, VOLUMES I AND II

Bonds and Bands
in Semiconductors

J. C. PHILLIPS

Bell Laboratories
Murray Hill, New Jersey

ACADEMIC PRESS   New York and London   1973

A Subsidiary of Harcourt Brace Jovanovich, Publishers

ACADEMIC PRESS, INC.
111 Fifth Avenue, New York, New York 10003

United Kingdom Edition published by
ACADEMIC PRESS, INC. (LONDON) LTD.
24/28 Oval Road, London NW1

Library of Congress Cataloging in Publication Data

Phillips, James Charles, DATE
        Bonds and bands in semiconductors.

        (Materials science and technology)
        Includes bibliographies.
        1.   Semiconductors.   2.   Energy-band theory of
solids.   3.   Free electron theory of metals.
4.   Chemical bonds.   I.   Title.
QC611.P58   537.6′22   72-12184
ISBN 0−12−553350−0

Contents

4 Lattice Vibrations   *77*

5 Energy Bands   *98*

6 Pseudopotentials and Charge Densities   *126*

7   Fundamental Optical Spectra   *154*

8   Thermochemistry of Semiconductors   *184*

9 Impurities   *219*

10 Barriers, Junctions, and Devices   *249*

Preface

Semiconductor technology has evolved rapidly since the invention of the transistor in the late 1940s by Bardeen, Brattain, and Shockley. With this progress has come a host of experiments and theories of a fundamental character concerning the electronic structure of covalent semiconductor materials. These developments have made it possible to understand the microscopic electronic structure of this family of materials more precisely and more systematically than that of any other assembly of interacting atoms. These advances can be described without the use of complicated mathematics or elaborate models. The description combines the language of solid state physics with that of chemistry and metallurgy. The utility of these approaches and their interrelations are apparent from the great extent to which the subject has developed.

The level at which this book has been written is typical of that of many introductory books on solid state physics. The approach, however, is interdisciplinary because many results are described both in terms of the energy bands of the physicist and the covalent bonds of the chemist. There is less emphasis on mathematical derivations than on relations between structure and properties. In these respects the reader will find emphasized here many basic properties of materials that are often ignored or regarded as accidents of nature in the traditional approach.

In writing a book such as this, one cannot avoid taking a simplified point of view which neglects small refinements contributed by many workers over the years. In any area of this large subject there is also the possibility of serious omissions, for which I beg in advance my colleagues' pardon. I should also mention that much of the work described here has benefited especially from the contributions of M. L. Cohen and J. A. Van Vechten.

1

Crystal Structures

WHAT IS A SEMICONDUCTOR?

The modern age of crystal electronics is based upon materials which are neither metals nor insulators. Such materials are called semiconductors, and their electrical properties are intermediate between those of metals and insulators. This is because of a rather special arrangement of the energy levels of electrons in semiconductors.

In isolated atoms electrons are found in definite states separated by discrete quanta of energy. In the hydrogen atom, the one electron which is bound to the proton may exist in a number of different states.

The probability distribution for the electron in a state α is given by

$$P_\alpha(\mathbf{r}) = |\psi_\alpha(\mathbf{r})|^2, \tag{1.1}$$

where $\psi_\alpha(\mathbf{r})$ is the wave amplitude. The energy of state α is represented by E_α.

ENERGY BANDS

Now consider an assembly of N atoms, each with its discrete energy levels $E_\alpha(i)$, where $i = 1, \ldots, N$ labels the atoms, and bring them together to form a crystal. Let ΔE_α [which is of order one electron volt (eV)] represent the average spacing of the atomic levels. As the atoms are brought together, the electrons originally bound to the attractive potential of one nucleus begin to interact with the potentials of other nuclei. This shifts each energy level E_α and eventually the formerly discrete levels become bands of levels with an energy spacing of order

$$\delta = \Delta E_\alpha / N. \tag{1.2}$$

This spacing δ becomes negligible when it is much smaller than kT, i.e., when

$$\delta \ll kT \sim \tfrac{1}{40} \text{ eV} \tag{1.3}$$

at room temperature. Thus a collection of several hundred atoms at solid densities already behaves in many respects like a solid. For practical purposes each group of energy levels forms a band of continuous levels in the solid state.

How are the electrons distributed in these energy levels? At zero temperature in a nonmagnetic crystal, the electrons fill the lowest available orbital energy levels twice with electrons of opposite spin, according to the Pauli exclusion principle. At higher temperatures the electrons are distributed among the lower energy levels indexed by n with probability

$$P_n = \frac{2}{\exp\left[(E_n - E_\mathrm{F})/kT\right] + 1}, \tag{1.4}$$

where E_F is the Fermi energy. Ordinarily $E_\mathrm{F} \gg kT$; then (1.4) corresponds to a very slightly broadened step function which is equal to two for $E_n \ll E_\mathrm{F}$ and to zero for $E_n \gg E_\mathrm{F}$.

METALS, INSULATORS, AND SEMICONDUCTORS

A metal is now defined as a material in which in the pure state E_F lies in a region of nonzero density of electronic band states

$$\rho(E) = dN(E)/dE, \tag{1.5}$$

where $N(E)$ is the number of electronic states per unit volume. Thus for a metal $\rho(E_\mathrm{F}) \neq 0$. In an insulator, on the other hand, there is a large energy gap between the highest occupied electronic state in the pure crystal at

$T = 0$, and the lowest unoccupied one. The Fermi energy E_F in the pure crystal falls approximately in the middle of this energy gap. Alkali halides are typical insulators, and their energy gaps are of order 5–10 eV.

Semiconductors are, qualitatively speaking, insulators with energy gaps of one to two eV or less. [If the energy gap is exactly zero, but $\rho(E_F) = 0$, so that the density of states increases linearly with $|E - E_F|$, one has a so-called semimetal. These behave like semiconductors in some respects]. Thus the energy gaps in the two most common semiconductors, Si and Ge, are 1.1 and 0.6 eV, respectively.

ALLOWED AND FORBIDDEN ENERGIES

In the valence energy region where atomic states overlap strongly, there is no reason a priori why $\rho(E)$ should ever be zero, and thus one might expect to find that all solids are metals. However, electrons in crystals satisfy the Schrödinger wave equation and the atoms are arranged periodically. One may say that the regions of energy where $\rho(E) = 0$ are forbidden to electrons, and those where $\rho(E) \neq 0$ are allowed (see Fig. 1.1). A

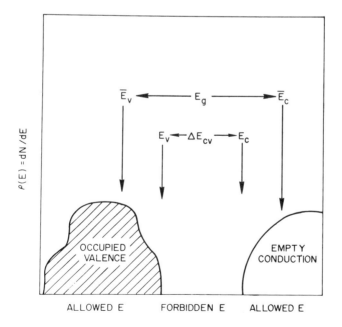

Fig. 1.1. Allowed and forbidden ranges of energy for electrons in a crystal. Also shown are valence and conduction bands and the energy gap between them (ΔE_{cv}) as well as the average gap E_g.

classical explanation of these allowed and forbidden regions can be given in terms of electromechanical analogies with coaxial cables and other circuits containing periodically repeated electrical elements. Solutions of the electromagnetic wave equation for these periodic arrays show pass and stop bands corresponding to zero and nonzero attentuation of propagating electromagnetic waves.

Mechanical waves passing along a periodically repeated array of masses and springs show similar behavior. These solutions are analogous to the allowed and forbidden energy ranges of electrons in crystals. A coaxial cable, however, is a one-dimensional system. In a three-dimensional crystal, one finds alternating allowed and forbidden energy ranges for electron waves propagating in any given direction. When one compares the allowed and forbidden energy ranges for different directions, however, it becomes apparent that something special must happen in order to make the forbidden energy ranges for different directions coincide, thus giving rise to an energy gap in all directions simultaneously. Typically the energy range of the occupied valence states shown in Fig. 1.1 is of order 10 eV, compared to a completely forbidden energy range of order 1 eV. This makes it even clearer that the coincidence of the forbidden energy ranges for wave propagation in different directions is rather improbable for semiconductors. For insulators, on the other hand, the forbidden energy range is typically as great or greater than the energy range of occupied valence states. This makes the insulator very stable with respect to chemical, pressure, or temperature changes. It explains why most naturally occurring minerals (like NaCl and Al_2O_3) have large energy gaps and insulating electrical properties. It also makes these minerals in general of less value to the physicist who is interested in manipulating their properties, for their very stability is itself an obstacle in this respect.

A crystal is a periodically repeated array of atoms, and the electron waves passing through this array in a certain direction should, according to electromechanical analogies, exhibit stop and pass bands (forbidden and allowed ranges of energy). But as we have seen, both coincidences of the forbidden energy region in all directions, and the presence of the right number of valence electrons in just such a way that electrons fill the allowed bands up to the forbidden energy region, but no further, are mathematically improbable. There are, however, many thousands of crystals which are known to be stable, and it could be that these coincidences occur in a few of them simply by accident. (There are physical phenomena, such as the antiferromagnetism of chromium, which actually are accidents of this nature.) This is not true, however, of semiconductors, for a large number of these are known, and they are all confined to crystals of closely related structure and chemical composition.

In molecules it is known that the spatial arrangement of atoms is often determined by the fact that the energy of the molecule is lowest when there are covalent bonds between the atoms. One may also apply this chemical picture to explain the crystal structure of semiconductors. For some people electromechanical analogies can be quite helpful, but because there are so many semiconductors with related properties the chemical view is usually preferred. This is because all classical analogies suffer from the weakness that the basic parameters of the electromechanical model appear arbitrary.

VALENCE BONDS

From chemical experience with molecules it has been learned that trends in electronic parameters (such as the energy gap or forbidden energy range ΔE_{cv} in Fig. 1.1) can often be explained in bond language. Because the bonding properties depend on the atomic potentials, the chemical explanation represents a condensed description of what one finds from solving the Schrödinger wave equation for electrons in the crystal. Because it discards many fine points in order to focus on important trends, the chemical approach is less rigorous mathematically but more intuitive. We use it to describe in a simplified way the three-dimensional energy bands of semiconductors and how these vary from one material to another. When necessary for specific details we can always fall back on energy bands calculated from solving the Schrödinger equation, but for many purposes it is sufficient to know that such bands are obtainable in principle.

BOND COUNTING

In chemical theory it is not always obvious just how many valence electrons are actually needed to saturate the covalent bonds between atoms, or, stated differently, what the relationship is between atomic coordination numbers and chemical valence. This problem is resolved in energy band theory by requiring that in the pure crystal there be just enough valence electrons to fill all the valence bands (which lie below the energy gap) and none left over for the conduction bands (which lie above the energy gap; see Fig. 1.1). Thus from the band viewpoint bonds are merely a convenient mnemonic description for band filling. However, as G. N. Lewis showed (around 1915), the counting of valence electrons in molecules by using two dots to represent each bond, corresponding to two electrons with opposing spins, can be quite useful. (For example, Lewis would represent an A–B bond by A : B.) Later Pauling found it necessary to generalize these ideas [1] to the higher coordination numbers commonly found in crystals by assuming that the actual bonds, when fewer in number than the coordination number, "resonated" from one position to another. In certain structures, such as white Sn, it is not easy to determine the number of resonating bonds.

ATOMIC ORBITALS

In spite of these difficulties the fact remains that there are many similarities between molecular and crystalline structures which are best described in bonding language. These can be expressed in terms of atomic orbitals $\varphi_{nl}(\alpha)$ around each atom α. The subscript n labels the principal quantum number of the atomic orbital, and l its orbital angular momentum, with $l = 0$ corresponding to s states, $l = 1$ to p states, and $l = 2$ to d states. There are $2l + 1$ independent orbitals for each l, and the important valence orbitals are usually the s and p ones, with occasionally a small admixture of d states. The s and p states alone give a maximum of $1 + 3 = 4$ valence states for each value of n, and generally we are concerned with the largest value of n occupied by electrons in the neutral atom. (Electrons with smaller values of n belong to the atomic core, while states with larger values are unlikely to be occupied.)

The atomic valence configurations of carbon, silicon and germanium are:

$$C: (\text{core}) + 2s^2 2p^2, \tag{1.6}$$

$$Si: (\text{core}) + 3s^2 3p^2, \tag{1.7}$$

$$Ge: (\text{core}) + 4s^2 4p^2. \tag{1.8}$$

When these atoms form covalent bonds in the crystal, it is no longer appropriate to discuss their electronic configurations in terms of atomic orbitals. An improvement which is sufficient for the moment is to introduce a new set of states directed from one atom towards its nearest neighbors.

HYBRIDIZED ORBITALS

A hybridized orbital χ_n is a linear combination of atomic orbitals φ_{nl}. Thus for carbon for example,

$$\varphi_{2s}(\mathbf{r}) = f_{2s}(r), \tag{1.9}$$

$$\varphi_{2p}(\mathbf{r}) = r^{-1}(x, y \quad \text{or} \quad z)f_{2p}(r). \tag{1.10}$$

Contours of constant probability or charge density for 2s or 2p electrons of carbon are sketched in Fig. 1.2. A hybridized orbital mixes the s and p atomic orbitals. For example, the hybridized orbital χ_{sp} is

$$\chi_{sp}(\mathbf{r}) = f_{2s}(r) + (z/r)f_{2p}(r). \tag{1.11}$$

A hybridized orbital is a *superposition* of two atomic amplitudes, and the associated probability density will exhibit interference effects. Thus (1.11) exhibits a maximum amplitude near the atomic radius $r = r_a$ when $\cos \theta = z/r = 1$, i.e., in the direction of the $+z$ axis. This means that χ_{sp} could be used to describe a bond between C and an atom along the z axis.

There are a number of possible linear combinations that can be made of atomic orbitals to form hybridized orbitals. Which will describe the bonded state best? Consider, for example,

$$\chi_{sp^2} = f_{2s}(r) + [\pm(x/r) \pm (y/r)]f_{2p}(r), \tag{1.12}$$

$$\chi_{sp^3} = f_{2s}(r) + [\pm(x/r) \pm (y/r) \pm (z/r)]f_{2p}(r), \tag{1.13}$$

which point in the $(\pm1, \pm1, 0)$ and $(\pm1, \pm1, \pm1)$ directions, respectively. Level contours associated with these orbitals are shown in Fig. 1.2.

Strictly speaking, each χ_{sp^n} should be multiplied in (1.11)–(1.13) by a normalization factor $(1 + n)^{-1/2}$. The directed character of these orbitals can be used for qualitative arguments. For example, consider the variations in C–H bond lengths in hydrocarbon molecules. According to Fig. 1.2, the tetrahedral hybridized orbital sp³ projects farthest away from the nucleus, while the sp orbital projects least. This is in agreement with observed C–H bond lengths, which vary according to tetrahedral (sp³, C_2H_6) > trigonal (sp², C_2H_4) > digonal (sp, C_2H_2).

In general the atomic orbitals that are used to form hybridized bonding orbitals are not the same ones that are occupied in the ground state of the atom. For example, in diamond the ground valence configuration of the atom [Eq. (1.6)] is 2s²2p², whereas the hybridized configuration appropriate for the diamond crystal structure is 2s2p³. When more than one kind of atom is present in the crystal, there is usually some redistribution of electronic charge or charge flow from one atom to another, both being neutral in their original atomic states. In each case it usually costs energy of order 5–10 eV/atom to promote the electrons into the hybridized states. This energy is recovered in the crystal through interactions with adjacent atoms, and overall the total energy is lowered, usually by an amount of order 1 eV/electron. The cohesive energy of order 1 eV/electron is about the same regardless of whether the crystal is metallic, ionic, or covalent. The energy differences between covalent, ionic, or metallic structures are usually small compared to the cohesive energy and are of order 0.1 eV/elec-

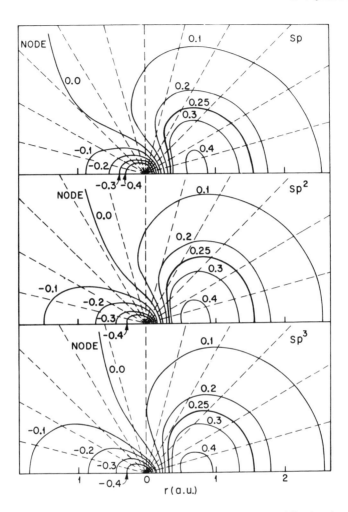

Fig. 1.2. Contours of constant wave function amplitude of hybridized carbon orbitals sp^n. The carbon covalent radius is about 1.5 Bohr units, and corresponds to the contour of amplitude 0.25, shown in the figures as heavier curves [from W. E. Moffitt and C. A. Coulson, *Phil. Mag.*, Ser. 7, **38**, 634, (1947)].

tron, so that predicting whether a given material will be covalent, ionic, or metallic requires a very accurate theory.

BONDING DEFINITIONS AND RULES

Directed valence bonds as shown in Fig. 1.2 can be used to define certain rules of covalent bonds. The important properties [2] of the directed orbitals can be expressed in terms of bonding, nonbonding, and anti-bonding orbitals:

i. A bonding orbital ψ_b consists of two directed orbitals $\chi_n(\alpha)$ and $\chi_n(\alpha')$ associated with nearest-neighbor atoms α and α' combined in phase in such a way that ψ_b is large in the bonding region between the atoms, as shown in Fig. 1.3..

ii. An antibonding orbital ψ_a is similar to a bonding orbital except that the phase has been reversed between atoms. Thus ψ_a has a node in the bonding region.

iii. A nonbonding orbital is usually centered on one atom and has little directional character.

The rules for bonds are:

1. The number of independent hybridized orbitals χ_n is equal to the number of atomic orbitals φ_{nl}.

2. In order of increasing energy one has bonding, nonbonding, and anti-bonding states. Usually the first two groups are occupied, and the third group is not. The bonding electrons have lowest energy because they

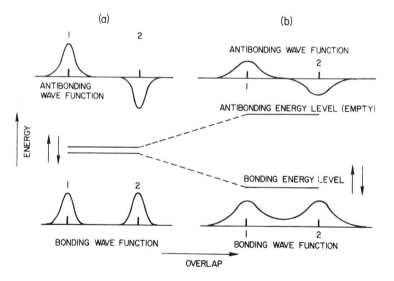

Fig. 1.3. Formation of bonding and antibonding states on going from atom (a) to crystal (b). In bonding states the wave function amplitude ψ_b associated with a directed orbital centered on atom 1 interferes constructively or in phase with a similar orbital centered on atom 2 in the bonding region between atoms 1 and 2. Similarly, the amplitude ψ_a of the antibonding orbital exhibits a node in the bonding region.

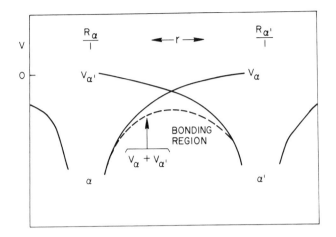

Fig. 1.4. Overlap of ion core potentials V_α and $V_{\alpha'}$ in the region between α and α' makes this region especially attractive to electrons in orbitals centered on α or α'.

benefit most from the large attractive potential produced by overlap of the Coulomb potentials of the positively charged ion cores α and α'. This is shown in Fig. 1.4.

3. Suppose that the bonding orbital has the form

$$\psi_b = \chi_m(\gamma) + \lambda\chi_n(\alpha),\tag{1.14}$$

where $\chi_m(\gamma)$ and $\chi_n(\alpha)$ are amplitudes such that the probability density of the bonding orbital ψ_b, given by (1.1), is

$$P_b(\mathbf{r}) = |\psi_b(\mathbf{r})|^2 = |\chi_m(\mathbf{r} - \mathbf{R}_\gamma)|^2 + \lambda^2 |\chi_n(\mathbf{r} - \mathbf{R}_\alpha)|^2.\tag{1.15}$$

The parameter λ is a weighting factor. By suitably normalizing χ_m and χ_n, one can say that the bonding electron spends a fraction f_γ of its time on the γ atom and a fraction f_α on the α atom, where

$$f_\gamma{}^b = 1/(1 + \lambda^2),\tag{1.16}$$

$$f_\alpha{}^b = \lambda^2/(1 + \lambda^2).\tag{1.17}$$

There are several possible values for λ. If the atoms γ and α are equivalent (examples are Si and Ge, see below), then $\lambda = 1$. If γ is the cation and α the anion, then $\lambda > 1$.

The conventional valence picture is based upon an antibonding state of the form

$$\psi_a = -\lambda\chi_m(\gamma) + \chi_n(\alpha).\tag{1.18}$$

This gives rise to the nodal behavior pictured in Fig. 1.3. It also predicts that

$$f_\gamma{}^a = f_\alpha{}^b, \tag{1.19}$$

$$f_\alpha{}^a = f_\gamma{}^b. \tag{1.20}$$

This means that if the bonding states spend a greater fraction of the time on the anions, the antibonding states will spend a greater fraction of the time on cations.

4. Bonding and antibonding states only are found in compounds with an average of four or fewer valence electrons per atom. In electron-rich compounds with an average of more than four valence electrons per atom, nonbonding states are occupied as well. The differences in energy between nonbonding orbitals on cations and anions are significant.

BOND ENERGY GAPS AND BAND ENERGY GAPS

One of the most important differences between the bond picture and the band picture for semiconductors, and one which has often been confused in the literature, is the energy difference E_g between the bonding and antibonding energy levels, and the energy difference ΔE_{cv} between E_v, the top of the valence band, and E_c, the bottom of the conduction band. It is ΔE_{cv} that determines the electrical properties of the semiconductor, and E_g that most influences structural properties. We shall see later how to define E_g, but for the moment it is sufficient to notice that ΔE_{cv} may be much less than E_g. This is because

$$E_g = \bar{E}_c - \bar{E}_v, \tag{1.21}$$

where $\bar{E}_c$ and $\bar{E}_v$ are *average* conduction and valence band energies, respectively. By definition $E_v > \bar{E}_v$, $E_c < \bar{E}_c$, so that (as illustrated in Fig. 1.1).

$$\Delta E_{cv} < \text{ or } \ll E_g. \tag{1.22}$$

For example, in Si and Ge, E_g is about 4.4 eV, compared to $\Delta E_{cv} = 1.1$ and 0.7 eV, respectively. Further examples are given in Table 1.1.

TETRAHEDRAL COORDINATION

In covalent structures the directed orbitals are oriented towards the nearest-neighbor atoms, thereby lowering the energy of occupied bonding orbitals. Most semiconductors have structures in which each atom is tetrahedrally coordinated, corresponding to the ψ_{sp^3} orbitals (1.13) with the product of the three $\pm$ signs positive on one atom and negative on the

TABLE 1.1

Average Bond and Minimum Energy Gap

Crystal	ΔE_{cv} (eV)	E_g (eV)
Diamond	5.7	13.5
Si	1.1	4.8
Ge	0.7	4.3
Gray Sn	0.0	3.1
InSb	0.2	3.7
GaAs	1.4	5.2
ZnSe	2.3	7.0
ZnO	3.6	12.0

other. An example is the diamond lattice (shown in Fig. 1.5), which consists of two interpenetrating face-centered cubic lattices. Thus each unit cell contains two atoms, and the unit cells are close-packed. But each atom has only four nearest-neighbors, in contrast to a close-packed monatomic

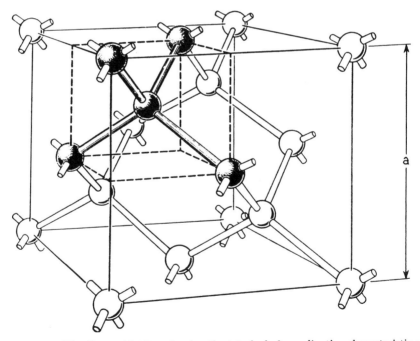

Fig. 1.5. The diamond lattice, showing the tetrahedral coordination characteristic of sp³ bonds [from W. Shockley, "Electrons and Holes in Semiconductors." Van Nostrand Reinhold, Princeton, New Jersey, 1950].

metal with its 12 nearest-neighbors. When the interpenetrating lattices are hexagonal close-packed, each atom still has four nearest-neighbors and 12 second-nearest ones, but each unit cell contains four atoms and the crystal has an axis of hexagonal (six fold) symmetry. The latter structure is called wurtzite. It occurs with more ionic semiconductors such as CdS or ZnO. When the two interpenetrating face-centered cubic lattices contain different atoms (such as GaAs) the structure is called zincblende, or more properly sphalerite. (Zincblende is the chemical term for ZnS, which occurs in both sphalerite and wurtzite structures).

According to the classical valence picture, all the elements from column IV of the periodic table (C, Si, Ge, Sn, and Pb) should be capable of forming diamond-type crystals with the structure shown in Fig. 1.5. Each atom has four valence electrons which occupy the tetrahedrally hybridized sp^3 orbitals pointing towards the four nearest neighbors. This is true for C through Sn, but Pb is found only in a close-packed metallic structure. In general one finds that with increasing core size (increasing atomic number) there is a tendency to pass from covalent to metallic binding.

A number of binary compounds $A^N B^{8-N}$ with eight valence electrons per atom pair are also found in tetrahedrally coordinated structures. This suggests that (at least in the simpler covalent structures where each atom has the same number of nearest neighbors), it is the *average* number $\bar{N}$ of valence electrons per atom which is important. Using $\bar{N}$ one can formulate the rule that for such structures the coordination number is $8-\bar{N}$. This "octet" rule describes a large number of systems, as shown in Table 1.2.

TABLE 1.2

Structures, Group Number, and Coordination Numbers of
Elemental Covalent Crystals

Element	Group number, N	Coordination number, $8 - N$	Coordination configuration
C, Si, Ge⎫ ⎬ α-Sn ⎭	4	4	Three-dimensional tetrahedral network
White P	5	3	P_4 tetrahedra
Black P, As⎫ Sb, Bi ⎭	.5	3	Puckered double layers
S, Se, Te	6	2	Rings, chains
Br, I	7	1	Pairs

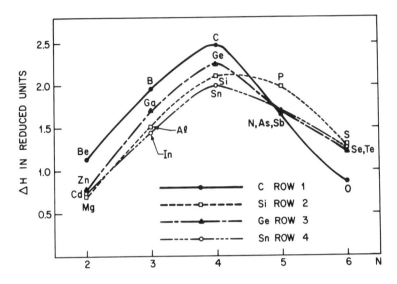

Fig. 1.6. Heats of atomization of solid elements at room temperature or at the melting point, whichever is lower. The heats for each element A in each row α are given in reduced units, $\Delta H(A)/\Delta H(\alpha)$, where $\Delta H(\alpha)$ is the average of ΔH for the elements with $N = 2$ and $N = 6$ of the row of the periodic table to which element A belongs.

The octet rule shows that valence bonds determine coordination numbers for $N \geq 4$. One would therefore expect cohesive energies of elemental crystals to peak at $N = 4$, where the number of valence bonds is largest. (For $N \leq 4$, metallic bonding predominates, and this is weaker than covalent bonding). In Fig. 1.6 the heats of atomization at the melting point of the elements from the first through fourth rows are shown, plotted in reduced units. (The energies are reduced by dividing the energy for an element in a given row by the average energy for the $N = 2$ and $N = 6$ elements of the same row, in order to emphasize the similar trends in each row as a function of N). The curves all peak at $N = 4$, with a value about twice that of the average of the $N = 2$ and $N = 6$ elements. This shows that for each row one may say that the energy per bond is nearly constant. The great stability of the diamond-type crystals comes about because for $N = 4$ the smaller of N and 8–N reaches a maximum value.

LAYER STRUCTURES

In addition to the tetrahedral sp^3 structures, there is one other basic type of semiconductor structure, which utilizes sp^2 bonds. These bonds lie in the xy plane and give rise to layer structures, usually of hexagonal symmetry (see Fig. 1.7). An example is graphite, a layer form of carbon. In each layer the carbon atoms form interlocking hexagonal rings similar to the benzene ring. The sp^2 bonds are shorter than (C–C, sp^3) bonds, 1.42 Å compared to 1.54 Å, in accord with Fig. 1.2. The C–C distance between adjacent C atoms in neighboring planes is 3.40 Å. The sp^2 bonds are referred to as σ bonds because they are even with respect to reflection in the plane containing them. The p_z orbitals are referred to as π orbitals. Different layers are well-separated, so that the binding between them is not covalent. The atoms in one layer are situated above the vacant centers of the rings in the layer below. The binary compound BN also forms hexagonal layers, but in that case the boron atoms of one layer are situated directly above the N atoms. This arrangement is favored by the Coulomb attraction between the B and N atoms.

Tetrahedral and layer structures represent extremes of nearly isotropic material on one hand and nearly two-dimensional properties on the other. For example, although wurtzite crystals have hexagonal symmetry, because of nearly tetrahedral nearest-neighbor configurations in most cases they are nearly isotropic in their physical properties; e.g., the index of refraction for light polarized along the hexagonal axis usually differs by a few percent or less from that of light polarized perpendicular to the axis.

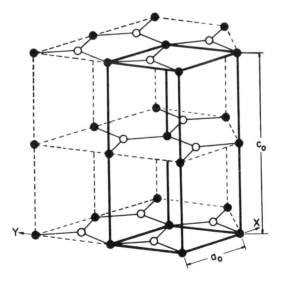

Fig. 1.7. The arrangement of atoms in the graphite structure. Both the filled and empty circles are lattice sites. The hexagonal unit cell is indicated by the heavy lines. [from R. W. G. Wyckoff, *Crystal Structures*. Wiley (Interscience), New York, 1965].

Layer crystals, on the other hand, frequently exhibit anisotropy of as much as 50% in their indices of refraction. Free carriers are often effectively confined to a single layer or set of layers, and have high mobility for electric fields parallel to a layer plane, but very low mobility for electric fields normal to the layer planes.

Most semiconductor structures are related to either the tetrahedral diamond or layer graphite structures. In order for the directed valence orbitals to be fully occupied, it is usually necessary that the average number of s–p valence electrons per atom be four. In a binary crystal this means a structural formula $A^N B^{8-N}$, which gives rise to group IV crystals (diamond, Si, Ge, and gray Sn), or III–V zincblende crystals, or II–VI and some I–VII crystals which may have either zincblende or wurtzite structures. Tetrahedrally coordinated ternary crystals of formula $A^I B^{III} C_2^{VI}$ and $A^{II} B^{IV} C_2^V$ are known. These structures are stabilized by both covalent bonding and by the close-packing of the C anions, which are the centers of most of the valence charge.

FLUORITE BONDS

An exception to the rule that sp^3 bonding requires an average of four electrons per atom is provided by crystals in the fluorite structure with formula RX_2. These contain either eight electrons per formula unit ($A_2^{II} B^{IV}$) or 16 ($A^{II} B_2^{VII}$ or $A^{IV} B_2^{VI}$). An example of the former is Mg_2Si, of the latter CdF_2 or $BaCl_2$. The former are covalent and exhibit optical spectra similar to Si and Ge, while the latter tend to be ionic like NaCl. An interesting intermediate case is GeO_2, which exhibits some covalent and some ionic properties. Each X atom has about it a tetrahedron of R atoms, so that the Mg atoms in Mg_2Si are sp^3-hybridized. The X atoms form a simple cubic lattice.

RELATIVISTIC STRUCTURES

The rule that s and p orbitals hybridize to form directed bonds is usually valid when the energy gap E_g is sufficiently large compared to the difference $E_{np} - E_{ns}$ between the np and ns valence electrons of the free atom. As n increases, all these energies decrease, but as we shall see in the next chapter, E_g decreases more rapidly than E_{np} and E_{ns}. Moreover, for $n = 4$ and $n = 5$ relativistic effects reduce E_{ns} compared to E_{np}. This is because in the neighborhood of the nucleus, where ψ_s remains nonzero as ψ_p tends to zero, $E = V(r)$ may be comparable to $2mc^2$. This leads through relativistic mass

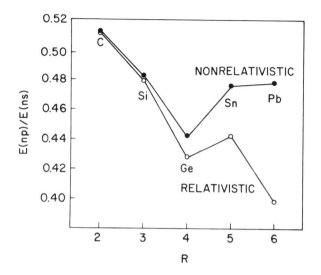

Fig. 1.8. The decrease in magnitude of np energies compared to ns energies for group IV elements both omitting and including relativistic mass enhancement of s electrons near the nucleus, as a function of row number R.

enhancement to an increase in the binding energy of s states only because of a decrease in kinetic energy. It is the relativistic cross terms between **p** and **r** that give rise to the spin–orbit interaction $\lambda \mathbf{L} \cdot \mathbf{s}$, and for 4p and 5p valence states λ is of order 1–2 eV, and is comparable to E_g. These remarks are illustrated in Fig. 1.8 for the group IV elements, where $\alpha = E_{np}/E_{ns}$ is plotted against n. The energy values are taken from atomic calculations [3], with and without relativistic corrections. The change in α from $n = 2$ to $n = 5$ is increased by a factor of four by the relativistic corrections.

Relativistic effects then increase $(E_{np} - E_{ns})/E_{ns}$ and cause dehybridization. As a rule this weakens or destroys the energy gap between bonding $(\text{sp}^3)^+$ states and antibonding $(\text{sp}^3)^-$ states and for large n crystals become metallic, as in the case of Pb. An exception is the family of binary compounds $A^{IV}B^{VI}$ which contains ten electrons per atom pair, including PbS, PbSe, and PbTe. The closely related crystals A_2^V with A = As, Sb, or Bi are not semiconductors, but are semimetals. Partial ionicity helps therefore to stabilize the $A^{IV}B^{VI}$ crystals which form the NaCl or rock salt structure, with six nearest-neighbors situated along Cartesian axes. The level scheme is $(\text{s}^2)^\pm (\text{p}^3)^+$; i.e., both bonding and antibonding s levels are filled, together with the p_x, p_y, and p_z bonding levels, which form directed orbitals oriented towards the six nearest-neighbors. One may also say that the $(\text{s}^2)^\pm$ orbitals are nonbonding.

CHALCOGENIDES

The column VI elements excluding O are called chalcogenides. The elemental crystals S, Se, and Te are semiconductors with a crystal structure consisting of spiral chains parallel to the z axis. The six electrons per atom are distributed in bonds according to $(sp_z)^+(p_x)^\pm(p_y)^\pm$; i.e., only the antibonding orbitals $(sp_z)^-$ along the chains are empty. In this case one may say that the orbitals perpendicular to the chain axis, i.e., the $(p_x)^\pm$ and $(p_y)^\pm$ orbitals, are nonbonding. See Fig. 1.9.

DEFECT AND EXCESS COMPOUNDS

Some binary semiconductors are known with nearly tetrahedral, fluorite, or layer structures with fewer or more than four valence electrons per atom, e.g., Cd_3As_2 ($A_3^{II}B_2^V$), In_2Te_3 ($A_2^{III}B_3^{VI}$), and Bi_2Te_3 ($A_2^VB_3^{VI}$). The struc-

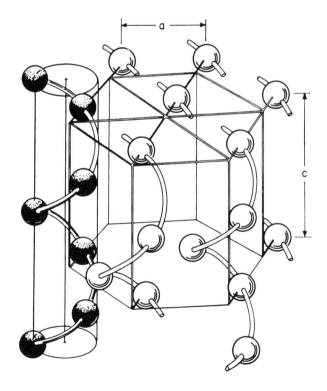

Fig. 1.9. The spiral structures of S, Se, and Te.

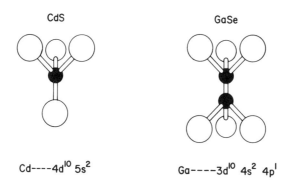

Fig. 1.10. In the crystal CdS the Cd metal atoms (solid spheres) are tetrahedrally coordinated with four S atoms, which is the zincblende configuration. In GaSe one has a sandwich structure, with planes of Se atoms separated by two planes of Ga atoms (shaded spheres). The metallic Ga atoms are tetrahedrally coordinated with three Se atoms and one Ga atom. The Se anions are on the outsides of the charge associated with nonbonding electrons.

tures tend to have one of the standard covalent forms, but because of too many or too few valence electrons, the structure is imperfect, with vacancies at certain sites and distortions of atoms in the vicinity of the vacancies. For example, Cd_3As_2 crystallizes in a structure resembling the fluorite structure, but with every fourth Cd site vacant. Electrostatic interactions then attract nearest-neighbor Cd ions towards Cd vacancies, shifting them by about 0.2 Å. Such imperfect structures generally make it difficult to grow large stoichiometric single crystals of the material.

Each electron band holds two electrons (of opposite spin) per unit cell. In $A^M B^N$ compounds with two atoms per unit cell this means that $M + N$ is an even integer. A complex structure which does *not* have an even number of valence electrons per binary pair is $A^{III}B^{VI}$, for example, GaSe. Each structural unit consists of four atoms arranged in layers BAAB. In this sandwich structure the anions form the outside layers (charge goes to the outside, as in a conducting sphere!). The A–B bonds are covalent, but the A–A bonds may be regarded as partly metallic or nonbonding. See Fig. 1.10 for a comparison of the sandwich structure with the tetrahedral zincblende structure.

TRANSITION METAL SEMICONDUCTORS

When some of the cations are transition metals a small energy gap may be found, e.g., in $CdMnS_4$, which has the spinel structure found in insulators. Other transition metal compounds have layer structures. These crystals are not semiconductors in the sense defined at the outset of this chapter, because the energy gap between the true valence and conduction bands is as large as in insulators of the rocksalt type. The observed small energy gaps involve the d levels of the transition metal atoms, either as initial or final states or both. Again it has so far proved difficult to prepare large single crystals of these materials.

Some representative crystals are described in bonding language in Table 1.3. Almost all the crystals of practical interest have the diamond, zincblende, wurtzite, or PbS structures. The remaining cases illustrate some of the variety of covalently bonded crystal structures [2].

BOND LENGTHS AND RADII

Crystallographers have long stressed the fact that for a given coordination number, or given state of valence hybridization, the bond length l_{AB}

TABLE 1.3

Crystal Structures and Hybridization States of Common Semiconductors

Crystal	Structure[a]	Hybridization
Diamond, Si	Diamond	$(sp^3)^+$
Ge, Gray Sn	Diamond	$(sp^3)^+$
GaAs, GaP	Zincblende	$(sp^3)^+$
InAs, InP	Zincblende	$(sp^3)^+$
BN, ZnTe	Zincblende	$(sp^3)^+$
ZnS	Zincblende	$(sp^3)^+$
ZnS	Wurtzite	$(sp^3)^+$
ZnSiP$_2$	Deformed zincblende	$(sp^3)^+$
Graphite	Hexagonal layers	$(sp^2)^+$
BN	Hexagonal layers	$(sp^2)^+$
Mg$_2$Si	Fluorite	$(sp^3)^+_{0.5}$
S, Se, Te	Spiral chains	$(sp_z)_2{}^+(p_x)_2{}^\pm(p_y{}^\pm)_2$
ZnP$_2$	Deformed zincblende	$(sp^3)^+$
Cd$_3$As$_2$	Imperfect fluorite	$(sp^3)^+_{0.8}$
In$_2$Te$_3$	Imperfect zincblende	$(sp^3)^+_{1.2}$
Bi$_2$Te$_3$	Imperfect zincblende	$(sp^3)^+_{1.3}$
PbS	Rocksalt	$(s^\pm)_2(p^3)^+$
GaSe	Sandwich	$(sp^2)^+(p_z)^+(p_z)^-_{0.5}$

[a] Deformed means a small uniaxial distortion is superimposed on the basic symmetry, e.g., to convert cubic to tetragonal. Imperfect means that at some lattice sites there are vacancies, which may attract or repel atoms at adjacent sites.

TABLE 1.4

Tetrahedral Covalent Radii in Å According to Pauling

	Be	B	C	N	O	F
	1.06	0.88	0.77	0.70	0.66	0.64
	Mg	Al	Si	P	S	Cl
	1.40	1.26	1.17	1.10	1.04	0.99
Cu	Zn	Ga	Ge	As	Se	Br
1.35	1.31	1.26	1.22	1.18	1.14	1.11
Ag	Cd	In	Sn	Sb	Te	I
1.52	1.48	1.44	1.40	1.36	1.32	1.28
	Hg	Tl	Pb	Bi		
	1.48	1.47	1.46	1.45		

between nearest-neighbors can be derived as the sum of radii r_A and r_B,

$$l_{AB} = r_A + r_B. \tag{1.23}$$

The definition (1.23) is not unique, for in an $A^N B^{8-N}$ crystal a constant ρ_N can be added to all $r_A{}^N$ and subtracted from $r_B{}^{8-N}$ for $N = 1, 2, 3$ without altering l_{AB}. For this reason different authors derive different values of r_A from the same values of l_{AB}.

To resolve this ambiguity one may attempt to find situations in which two atoms with valence N not equal to four are tetrahedrally coordinated and are bonded to each other. This determines r_A or r_B for atoms of that valence if one assumes that the atomic radius is independent of the number of nonbonding electrons. This is the method employed by Pauling.† His values of tetrahedral radii, which are widely used, are given in Table 1.4.

RATIONALIZED RADII

Since the completion of Pauling's work in the thirties there have been a number of studies of tetrahedral bonding in complex molecules where the average number of valence electrons per atom is greater than four. In general these show that even for tetrahedral coordination the atomic radii

† See [1, p. 246]. Strictly speaking, the S–S distance predicting by Pauling's table, which is 2.08 Å, is not in good agreement with the modern value for FeS_2, which is 2.14 Å. However, the S–S bond length in most of the crystalline forms of S is near 2.05–2.06 Å, in better agreement with Pauling's radii. The conclusion that Pauling's radii are determined by anion–anion lengths remains valid, because Fe–S interactions dominate in FeS_2.

are not constant, and instead decrease as the average charge on the atomic cores increases. That is presumably because the bonding electrons can take advantage of stronger attractive potentials associated with partially screened atomic cores in the bonding region between the atoms where the core potentials overlap strongly. The larger attractive potential makes possible greater localization of the bonding electrons and accordingly reduces the bond length.

One may confine oneself to the family of $A^N B^{8-N}$ crystals and ask for values of r_A and r_B when the average number of valence electrons per atom is four. In this case to a first approximation one finds that for a given row of the periodic table $r_A = r_B = r_4$, where r_4 is half the bond length of the elemental diamond-type crystal of that row. This implies that when A and B belong to different rows of the periodic table, one has approximately

$$l_{AB} = r_{4A} + r_{4B}, \qquad (1.24)$$

where r_{4A} is half the bond length of the diamond-type crystal of the row to which atom A belongs, and similarly for r_{4B}.

Exceptions to (1.24) are particularly large when A and B belong to the first row of the periodic table. For example, the bond length of BeO is about 7% larger than that of diamond. This expansion can be traced to the fact that the Be core 1s radius plus the O core 1s radius is more than twice the C 1s core radius; i.e., in the first row there is a substantial change in core radius as N changes. On this basis one can construct a table of rationalized tetrahedral atomic radii which involve besides the values of r_4 in diamond, Si, Ge and gray Sn only two parameters [4]. These radii are shown in Table 1.5. For each row they are more nearly constant than are Pauling's radii.

The conventional chemical valence picture stresses the valence electrons only, and de-emphasizes the roles played by the atomic cores. For tetra-

TABLE 1.5

Rationalized Tetrahedral Radii in Å

	Be	B	C	N	O	F
	0.975	0.853	0.774	0.719	0.678	0.672
	Mg	Al	Si	P	S	Cl
	1.301	1.230	1.173	1.128	1.127	1.127
Cu	Zn	Ga	Ge	As	Se	Br
1.225	1.225	1.225	1.225	1.225	1.225	1.225
Ag	Cd	In	Sn	Sb	Te	I
1.405	1.405	1.405	1.405	1.405	1.405	1.405

hedrally coordinated crystals this approach works best for compounds containing atoms from the Si, Ge, and Sn rows of the periodic table. Atoms from the Pb row do not hybridize well enough in most cases to form covalent structures, and prefer to form metallic structures, for the reasons discussed in connection with Fig. 1.8. On the other hand, atoms from the C row differ drastically in size from those of the Si, Ge, and Sn rows, as one can see from Tables 1.4 and 1.5. These size differences are important in many respects. For example, deviations from trends followed by compounds composed only of atoms from the Si, Ge, and Sn rows are found in $A^N B^{8-N}$ compounds containing A atoms from the Si, Ge, or Sn rows and B atoms from the C row (or vice versa). But if both atoms A and B belong to the C row, normal behavior is found.

These anomalies appear to correlate with sizable core corrections to atomic radii, the so-called core expansion effect [4]. Analysis of the atomic radii shows that this effect is definitely present in all the first-row atoms, but it is also present in the first three atoms of the Si row (Na, Mg, and Al), and deviations from normal covalent behavior are often found in compounds containing these elements. For example, in Table 3.4 the deviations from tetrahedral coordination (or ideal c/a ratio) are listed for a number of $A^{II} B^{IV} C_2^V$ compounds. By far the most anomalous is $MgSiP_2$, where a large deviation (0.232) is found even though all the atoms belong to the Si row. The corresponding compound from the Ge row, $ZnGeAs_2$, exhibits a deviation of only 0.034, roughly seven times smaller. It is thought that the Mg atom is the villain in $MgSiP_2$.

IMPURITY RADII

The values for covalent radii shown in Table 1.5 should be most accurate for treating sp^3-hybridized $A^N B^{8-N}$ crystals. They describe neutral atoms, so that one can use them to estimate strain fields associated with impurity atoms of the same valence as the atom they are replacing. When the impurity valence differs from that of the atom it replaces, an additional correction for Coulomb effects is necessary.

LAYER BONDS

For the less common case of sp^2 bonding only the planar bonds are covalent, and they are always shorter (stronger) than the sp^3 bonds. For example, the covalent radius of sp^3 carbon (diamond) is 0.77 Å, and that of sp^2 carbon (graphite) is 0.71 Å, substantially the same as benzene, 0.70 Å.

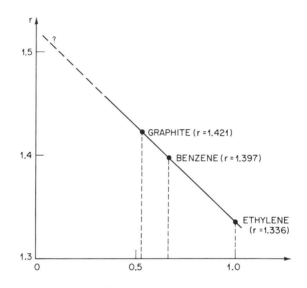

Fig. 1.11. The bond length in Å of carbon–carbon bonds against p, which is the total bond order -1; i.e., $p = 0$ corresponds to the ordinary tetrahedral single bond.

In organic chemistry [5] the sp^3 bond is called a single bond, while the graphite bond is said [5] to have a total bond order of 1.53, close to the benzene value of 1.67. Evidently multiple bonds are stronger than single bonds, and layer bonds are in general stronger within the layer than the corresponding tetrahedral bond would be.

From molecular studies it is known that certain elements are much more likely to form strong, short multiple bonds than others. Especially likely are the first-row elements C, N and O. Less affected but still possible are P and S. These elements in the liquid state (or "melt") will have lower energies as impurities in silicon or germanium than they would have been expected to have relative to the crystalline state, because of the formation of some multiple bonds in the liquid. One may denote the total bond order by $1 + p$ and plot bond length for A–B bonds against p. This is done in Fig. 1.11 for carbon-carbon bonds.†

SUMMARY

We have sketched here the structures of the more common and more useful semiconductors. The most prevalent structures are seen to be those

† See [5, p. 138]. Further examples of multiple bonds are C_2H_4, a double C–C bond with p = 1, represented by $H_2C{=\!=\!=}CH_2$, and C_2H_2, a triple bond with p = 2 represented by $HC{\equiv}CH$.

describable in terms of directed valence bonding orbitals which are doubly occupied by electrons of up and down spin. In the complete absence of interatomic overlap between directed orbitals centered on nearest neighbors, the bonding orbitals would have been degenerate in energy with the antibonding orbitals, as shown in Fig. 1.3(a). Overlap effects lower the energy of the bonding orbital and raise the energy of the antibonding orbital, as shown in Fig. 1.3(b).

With electrons only in the lower bonding orbital, the structure is stabilized by the energy gained from the formation of bonds. However, this energy must be compared with the energy required to form hybridized orbitals in the first place. For example, the ground valence state of C is $2s^2 2p^2$, and energy is required to promote an electron from the 2s to the 2p state in order to form hybridized $2s2p^3$ directed orbitals. Thus we need a quantitative measure of the energy gained from overlap effects as shown in Fig. 1.3. This is the subject of Chapter 2.

REFERENCES

1. L. Pauling, "The Nature of the Chemical Bond," Chapter 4. Cornell Univ. Press, New York, 1960.
2. F. Hulliger and E. Mooser, The Bond Description of Semiconductors: Polycompounds, *in* "Progress in Solid State Chemistry" (H. Reiss, ed.), Pergamon, Oxford, 1965.
3. F. Herman and S. Skillman, "Atomic Structure Calculations." Prentice-Hall, Englewood Cliffs, New Jersey, 1963.
4. J. A. VanVechten and J. C. Phillips, *Phys. Rev.* **2B**, 2160 (1970).
5. L. Salem, "The Molecular Orbital Theory of Conjugated Systems." Benjamin, New York, 1966.

2

Covalent and Ionic Bonds

In Chapter 1 the language of covalent bonds was used to give a qualitative description of the kinds of crystal structures in which the more common semiconductor crystals are found. This language can be made more quantitative by considering the specific family of binary tetrahedrally coordinated crystals of formula $A^N B^{8-N}$, that is, crystals with a total of eight valence electrons per atom pair. According to the language of valence bond theory used by Pauling and by Coulson, the valence bonds arise from overlapping tetrahedrally oriented sp³-hybridized valence orbitals. The properties of these orbitals which influence constitutive properties of the crystals can be described in terms of the potentials seen by the valence electrons localized near their respective ions. From studies of atomic energy levels and wave functions, we know that the potential seen by valence electrons near an ion of net charge $+Ze$ is given approximately by

$$-eV(r) = -Ze^2/r, \qquad r \gtrsim r_c, \qquad (2.1)$$

$$-eV(r) = 0, \qquad r \lesssim r_c, \qquad (2.2)$$

where r_c is the core radius of the ion. The precise meaning of (2.1) and

(2.2) will be explained in more detail later, but one can already see that it is the net charge Ze on the ion core, not the nuclear charge Ae (where A is the atomic number), which determines the properties of bonding orbitals. For each column of the periodic table, Z has approximately the same value, and for nontransition elements this is just the number of the column in the periodic table to which the atom belongs.

ELECTRONIC CONFIGURATIONS OF ATOMS

There are a great many semiconductor crystals. In order to keep their atomic structures conveniently in view an abridged form of the periodic table is presented for the reader's convenience in Fig. 2.1. The elements which form at least in one case, tetrahedrally coordinated $A^N B^{8-N}$ crystals, are marked with a double line, which is meant to signify the energy gap between bonding and antibonding states. In this figure the configurations of the valence and outer core electrons which affect the energy gap are also listed for each atom.

The ns and np valence electrons of course are the principal contributors

COLUMN / ROW	Iα	IIα	Ib	IIb	III	IV	V	VI	VII
1	3 Li $2s$			4 Be $2s^2$	5 B $2s^2 2p$	6 C $2s^2 2p^2$	7 N $2s^2 2p^3$	8 O $2s^2 2p^4$	9 F $2s^2 2p^5$
2	11 Na $3s$			12 Mg $3s^2$	13 Al $3s^2 3p$	14 Si $3s^2 3p^2$	15 P $3s^2 3p^3$	16 S $3s^2 3p^4$	17 Cl $3s^2 3p^5$
3	19 K $4s$	20 Ca $4s^2$	29 Cu $3d^{10}4s$	30 Zn $3d^{10}4s^2$	31 Ga $3d^{10}4s^2 4p$	32 Ge $3d^{10}4s^2 4p^2$	33 As $3d^{10}4s^2 4p^3$	34 Se $3d^{10}4s^2 4p^4$	35 Br $3d^{10}4s^2 4p^5$
4	37 Rb $5s$	38 Sr $5s^2$	47 Ag $4d^{10}5s$	48 Cd $4d^{10}5s^2$	49 In $4d^{10}5s^2 5p$	50 Sn $4d^{10}5s^2 5p^2$	51 Sb $4d^{10}5s^2 5p^3$	52 Te $4d^{10}5s^2 5p^4$	53 I $4d^{10}5s^2 5p^5$
5	55 Cs $6s$	56 Ba $6s^2$	79 Au $5d^{10}6s$	80 Hg $5d^{10}6s^2$	81 Tl $5d^{10}6s^2 6p$	82 Pb $5d^{10}6s^2 6p^2$	83 Bi $5d^{10}6s^2 6p^3$	84 Po $5d^{10}6s^2 6p^4$	85 At $5d^{10}6s^2 6p^5$

Fig. 2.1. Abridged version of the periodic table (excluding transition, rare gas, rare earth, and actinide elements, as well as hydrogen). The elements which form at least in one case tetrahedrally coordinated $A^N B^{8-N}$ crystals are marked with a double line, which is meant to signify the energy gap between bonding and antibonding states. The configurations of outer electrons in the ground atomic states are also indicated.

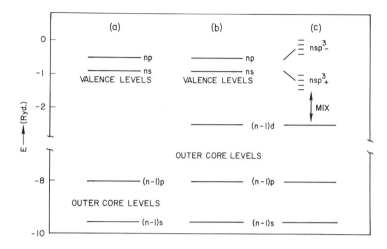

Fig. 2.2. Sketches of the atomic energies of outer electrons (a) in atoms not containing d core electrons, and (b) in atoms containing d core electrons. In (c) the mixing effects of overlapping atomic potentials in the crystal are indicated.

to the covalent bond. In Fig. 2.1, however, the outer $(n - 1)d^{10}$ electrons, which are usually assigned to the atomic cores, are listed as well. In most chemical discussions these electrons are ordinarily ignored. The d shell is full, and according to the exclusion principle these electrons can contribute to the bond only through promotion to unoccupied ns and np valence states. This process of promotion is weak because of the positions of energy levels shown in Fig. 2.2.

In (a) of this figure the energy levels of an atom from row 1 (the carbon row) or row 2 (the silicon row) are shown. The ns and np valence levels are bound by about Z^2/n^2 Ry, where Z is the net charge on the ion core. This is an energy E_v typically of order 1 Ry or less. The $(n - 1)s$ and $(n - 1)p$ electrons are much more strongly bound; their binding energies E_c are typically of order 10 Ry. These binding energies should be compared to the energy valence electrons can gain by interacting with the potentials of the ion cores of other atoms and other valence electrons. Call these interatomic interaction energies $\mathcal{I}$. They will have both kinetic and potential components. Kinetic energies of valence electrons in the solid are of order E_F, where E_F is the Fermi width of a free electron gas of density equal to that of the valence electrons. In terms of the number n_0 of valence electrons per unit volume one defines a Fermi wave number k_F and a Fermi energy E_F by

$$k_F = (3\pi^2 n_0)^{1/3}, \tag{2.3}$$

$$E_F = \hbar^2 k_F^2/2m, \tag{2.4}$$

where m is the electron mass. Potential energies are estimated by imagining the valence electrons to be rigidly displaced from the ions by a small amount and then allowed to oscillate (still as a rigid unit) about the original equilibrium position. The oscillation or plasma frequency is called ω_p and it is given by

$$\omega_p{}^2 = 4\pi n_0 e^2/m. \tag{2.5}$$

Potential energies associated with Coulomb interactions are of order of the plasma energy $\hbar\omega_p$. Values of ΔE_{cv} are usually much smaller than either $\hbar\omega_p$ or E_F, as one can see in Table 2.1. In general interaction energies $\mathcal{J}$ are of order E_g, which is of order (0.3–0.8) times E_F or $\hbar\omega_p$, depending on the ionicity of the crystal.

By comparing s and p core energies with the interaction energies $\mathcal{J}$ one can see that the change in s and p core states will be small on going from the free atom to the crystal. The situation is different, however, for the core d electrons. These are present in the heavier atoms from the third, fourth and fifth rows of the periodic table (Fig. 1.1).

CORE d ELECTRONS

The situation in the presence of an outer core $(n-1)d^{10}$ shell is shown in Fig. 2.2(b). Here the arrangement of energy levels is quite different. The value of $\Delta E = E_v - E_d$ is only about 1 Ry. As a result $(\mathcal{J}/\Delta E)^2$ is about 0.1. This is still rather small, but the differences between one semiconductor and another are also small. To understand these differences allowance must be made in the case of atoms from the Ge, Sn, and Pb rows for effects associated with the $(n-1)d^{10}$ outer core electrons. These mix with the valence levels as shown in (c) of Fig. 2.2.

TABLE 2.1

Plasma, Fermi, and Minimum Gap Energies

Crystal	$\hbar\omega_p$ (eV)	E_F (eV)	ΔE_{cv} (eV)
Diamond	31.2	28.9	5.7
Si	16.6	12.5	1.1
Ge	15.6	11.5	0.7
Gray Sn	12.7	8.7	0.0
InSb	12.7	8.8	0.2
GaAs	15.6	11.5	1.4
ZnSe	15.6	11.4	2.3
ZnO	21.5	17.6	3.6

The foregoing discussion summarizes most of what is known about the properties of atoms which influence the electronic structure of semiconductors. The periodic table, reflecting as it does the valence configurations of the atoms, is a valuable guide to understanding how atomic properties influence crystalline properties. By itself, however, the valence approach does not describe most of the important trends in semiconductor properties. The nature of the bonds themselves in the crystal is what is most important. For example, there is a great range of properties in $A^N B^{8-N}$ compounds for $N = 3$. One expects the compounds with $N = 2$ to be on the average more "ionic" than the ones with $N = 3$. This is correct, but it is also true that the most ionic compound with $N = 3$ is almost as ionic as the least ionic one with $N = 2$. By alloying, the physical properties of $A^N B^{8-N}$ semiconductors can be made to vary practically continuously. This behavior lies outside the usual valence picture, and to describe it one must examine the covalent bond itself more closely.

UNIVERSAL SEMICONDUCTOR MODEL

In the following discussion all tetrahedrally coordinated $A^N B^{8-N}$ semiconductors will be treated within the framework of a single model. Each crystal will be regarded as a particular case of the universal model which is described by particular values of the parameters of the model. To the extent that this can be done with accuracy, our picture of semiconductor structure will be greatly simplified. The success of this approach requires a careful choice of parameters which enter the model. The most important of these is the ionicity of the bond.

COVALENT AND IONIC CHARACTER

The ionicity of a bond is defined as the fraction f_i of ionic or heteropolar character in the bond compared to the fraction f_h of covalent or homopolar character. By definition these fractions satisfy the relation

$$f_h + f_i = 1.00. \tag{2.6}$$

In an elemental crystal like Si, one must have $f_h = 1.00$ and $f_i = 0.00$. On the other hand, we shall find that rock salt crystals like NaCl are more than 90% ionic. The more ionic crystals exhibit larger energy gaps between the valence and conduction bands. Because of this the more ionic crystals are less polarizable and they have correspondingly smaller dielectric constants.

SYMMETRIC AND ANTISYMMETRIC POTENTIALS

In a crystal of formula $A^N B^{8-N}$ we may construct a unit cell centered on an origin halfway between the two atoms. Denote by V_A and V_B the screened Coulomb potential seen by valence electrons outside the ion cores of atoms A and B, respectively. We assume that the potential of the crystal can be written as

$$V_{crystal} = \sum_\alpha V_A(\mathbf{r} - \mathbf{r}_\alpha) + \sum_\beta V_B(\mathbf{r} - \mathbf{r}_\beta), \qquad (2.7)$$

where $\mathbf{r}_\alpha$, $\mathbf{r}_\beta$ label the lattice sites of the sublattices of A, B atoms respectively. Looking at only one cell, and at only the potentials V_A and V_B associated with the atoms in that cell, we separate $V_{crystal}$ into the parts which are even and odd with respect to interchange of A and B by means of inversion about the origin. These we call $V_{covalent}$ and V_{ionic}, respectively:

$$V_{covalent} = V_A + V_B, \qquad (2.8)$$

$$V_{ionic} = V_A - V_B. \qquad (2.9)$$

At this point it should be obvious that f_h and f_i are somehow related to certain average values of (2.8) and (2.9) calculated with respect to certain wave functions of the crystal. The question is how to define these average values, and it is here that definitions diverge.

COULSON DEFINITION OF IONICITY

Coulson's school uses an explicit representation based on atomic orbitals [1]. Each valence wave function is written in the form

$$\psi_{valence} = \psi_{sp^3}(A) + \lambda\psi_{sp^3}(B), \qquad (2.10)$$

where $\psi_{sp^3}(A, B)$ denotes a hybridized valence orbital centered on atom A or B, respectively. The trial wave function (2.10) is inserted in the wave equation and an estimate is made of the total energy. When the latter is minimized, the best value of λ is obtained. Because the two atomic wave functions in (2.10) are assumed to be orthogonal, the ionicity defined by (Prob. = probability of finding a valence electron)

$$f_i = \frac{(\text{Prob. on A}) - (\text{Prob. on B})}{(\text{Prob. on A}) + (\text{Prob. on B})} \qquad (2.11)$$

is given simply by

$$f_i = (1 - \lambda^2)/(1 + \lambda^2). \qquad (2.12)$$

Within the framework of quantum theory based on atomic orbitals, equations (2.10)–(2.12) probably give about as good a definition of ionicity as can be obtained. The trouble with this definition is that it suffers from two weaknesses. First, the restriction to a trial function of the form (2.10) is much too severe, because the formation of the bond changes the localized orbitals about each atom from what they were in the free atom, making (2.10) a poor choice for defining that bond's ionicity. Second, the energy calculated does not give the cohesive energy of the crystal, or the observed lattice constant, so that varying λ to minimize that energy is a much less meaningful process than it might appear to be at first sight.

PAULING DEFINITION OF IONICITY

The latter difficulty was already recognized in the thirties. To avoid it Pauling based his definition [2] of ionicity not on the total energy of the bond but on empirical heats of formation. He noticed that in molecules the reaction

$$AA + BB \rightarrow 2AB \qquad (2.13)$$

is almost always exothermic when the A–B bond is partially covalent, i.e., the energy of an A–B bond is lower (more negative, larger in absolute magnitude D_{AB}) than the average absolute energies of A–A and B–B, i.e.,

$$D_{AB} > \tfrac{1}{2}(D_{AA} + D_{BB}). \qquad (2.14)$$

This difference is called Δ_{AB}, the extra-ionic energy, because the A–B bond is partially ionic, while the A–A and B–B bonds are strictly homopolar or covalent.

The explanation for the fact that $\Delta_{AB} > 0$ is the following. One of the atoms (say A) has greater power to attract electrons than the other. Denote the power of an atom A to attract electrons to itself by a dimensionless number called its electronegativity X_A. Expand Δ_{AB} in powers of $(X_A - X_B)$ and keep only the first nonvanishing term. By symmetry this gives to lowest order in $(X_A - X_B)$ the approximate result

$$\Delta_{AB} = (\text{constant with dimensions of energy})\,(X_A - X_B)^2. \quad (2.15)$$

One can imagine that a certain number of electrons proportional to $(X_A - X_B)$ has been transferred from atom B to atom A. The Coulomb interaction between the ionic charge left behind and the valence charge transferred is proportional to $(X_A - X_B)^2$, and this is the origin of the extra-ionic energy.

We see then that $X_A - X_B$ is a measure of ionicity. But by definition f_i

never exceeds one, and as $X_A - X_B$ becomes large f_i tends to one. Moreover ionicity of an A–B bond should be the same as ionicity of a B–A bond. This suggested to Pauling the definition [2] of ionicity of a single bond,

$$f_i = 1 - \exp\left[-(X_A - X_B)^2/4\right], \tag{2.16}$$

where the constant in (2.15) has been chosen so that X_A increases by 0.5 when Z changes by one in the first row elements. The choice of (2.16) makes the NaCl single bond about 67% ionic on Pauling's scale.

EXTENSION OF PAULING'S DEFINITION TO CRYSTALS

In most molecules the coordination number of an atom is less than or equal to its formal valence N, which is equal to Z for $Z \leq 4$ and to $8 - Z$ for $Z \geq 4$. In crystals, on the other hand, the coordination number generally exceeds N. In $A^N B^{8-N}$ crystals the coordination number is usually four or six, so it is only in diamond-type crystals, where N equals four and the coordination is tetrahedral, that one has single bonds.

To resolve this conflict Pauling introduced the concept of *resonating bonds*. In each compound there are N valence bonds per atom, and these are pictured as shared with all the four or six nearest neighbors of each atom. This means that the degree of covalency $f_h = 1 - f_i$ is also shared, so that the single-bond ionicity f_i must be replaced by a resonating-bond ionicity f_i' defined by

$$1 - f_i' = N(1 - f_i)/M, \tag{2.17}$$

where M is the number of nearest neighbors and f_i is defined by (2.16). One can rewrite (2.17) as

$$f_i' = 1 - N/M + Nf_i/M. \tag{2.18}$$

When $N < M$, $f_i' > 0$ even when $f_i = 0$.

LIMITATIONS OF PAULING'S DEFINITION

If one is concerned with heats of formation of diatomic molecules, then Pauling's equations (2.15) and (2.16) are probably as good a definition of ionicity as any. Later workers showed that Pauling's results could be improved somewhat by defining D_{AB} as the average bond energy of the polyatomic molecules AB_n or A_mB, where n is the valence of atom A or m is the valence of atom B, thus making certain that as many electrons as possible of A or B, respectively, are actually bonded.

This improvement does not remove the central weaknesses of Pauling's definition, which become apparent when we consider $A^N B^{8-N}$ semiconductors and equation (2.18). The semiconductors are much simpler than diatomic or even polyatomic molecules, because all valence electrons are bonded, and all bonds are equivalent. (E.g., in the molecule AB_n, the n valence electrons of A are all bonded, but only one of the valence electrons of each B atom is bonded). However, in the crystalline case D_{AB} refers to the heat of formation of the A–B crystal from A–A and B–B at STP (standard temperature and pressure). For example, the heat of formation of BN is measured relative to crystalline boron and gaseous nitrogen. The former has a complicated network structure which cannot be well described by hybridized orbitals, while the latter consists of sp-bonded diatomic molecules. These changes in structure make a significant contribution to D_{AB}. The second criticism of Pauling's definition is even more serious. The electronegativities $X_A - X_B$ in (2.15) correspond to some kind of average of $V_{ionic} = V_A - V_B$ as defined by (2.9). This ionic energy must compete with the homopolar energy associated with $V_{covalent} = V_A + V_B$ as defined by (2.8). The latter is the only other energy in the problem, and so it must be related to the prefactor in (2.15), treated as a constant by Pauling. In fact this "constant" varies by more than a factor of four from the first to fourth row of the periodic table. This generates inconsistencies in Pauling's estimates of ionicities. These can be eliminated when A and B belong to the same row of the periodic table by choosing X_A and X_B suitably. However, when A and B belong to different rows of the periodic table, some inconsistencies still remain.

Another problem in extending the molecular approach to crystals is the meaning of the valence N. For most nontransition atoms the choice $N = Z$ or $8 - Z$ works well. The IB atoms Cu, Ag, and Au often are divalent in molecules, however, which means that one of the $(n - 1)d^{10}$ core electrons is participating in the bonding along with the ns valence electrons. Thus in the silver halides Pauling suggested that N was probably equal to two. However, the halogen ions are *always* monovalent, so that one could equally well argue that N should be one. An ambiguity has arisen because the cation valence is one more than the anion valence. This makes the definition of N appear arbitrary, and leaves one in doubt about the meaning of equation (2.18).

THE MIDDLE WAY

Now that we have recognized the difficulties encountered by both theoretical and empirical approaches to defining ionicity, are we justified in abandoning the attempt altogether? No, because there is a middle way which combines the best features of both, and which is subject to the defects of neither. Finding and describing this middle way is one of the objects of this book.

As we have seen, the formation of covalent bonds leads to the appearance of an energy gap between bonding and antibonding states. We can estimate the average energy of an sp^3-hybridized state centered on atom A in terms of free-atom energies E_s and E_p as

$$E_A = [E_s(A) + 3E_p(A)]/4. \qquad (2.19)$$

For two atoms A and B, differences between E_A and E_B will lead to charge transfer when the two atoms are bonded together.

HOMOPOLAR ENERGY GAPS

When both atoms in the unit cell are the same, we have a completely homopolar bond and the hybridized orbitals have the same energy E_A. The potential V_{ionic} is zero in this case, and the degeneracy of the overlapping sp^3 orbitals is removed by interaction with $V_{covalent}$. This gives rise to the bonding and antibonding levels described by equations (1.13) and (1.17). The levels are separated in energy by the average homopolar energy gap which we denote by E_h. The latter energy represents a suitable average of the bonding effects of the homopolar potential $V_{covalent}$.

How is this average energy gap to be determined? Any method of measuring total energy will involve the energy gap per electron times the number of valence electrons per atom contributing to the bond energy. If this number is known accurately, then the microscopic energy gap can be obtained from a macroscopic energy. We choose to use optical polarizabilities to determine E_h (homopolar crystals) and later E_g (partially ionic crystals) because of the connection that exists between optical oscillator strengths and valence electron density n_0. This connection is analogous to the f-sum rule for atomic oscillator strengths and it will be discussed in detail in later chapters. For the moment we quote only the basic relation between $\epsilon(0)$, the square of the optical index of refraction in the limit $\omega \to 0$, and E_h. This relation is

$$\epsilon(0) = 1 + (\hbar\omega_p/E_h)^2 A. \qquad (2.20)$$

Here A is a number close to unity, which represents a small correction to the bond model due to the banding effects illustrated in Fig. 1.1. Note that the plasma frequency squared, ω_p^2, is proportional to n_0, according to (2.5). When only s and p valence electrons contribute to the bond, n_0 is equal to the average number of s–p valence electrons per atom, which of course is four in $A^N B^{8-N}$ compounds. When d electrons participate in the bonding, n_0 is determined optically from the f-sum rule.

In diamond and silicon there are no d core electrons, so that n_0 is indeed four per atom. The energy gap E_h should depend only on bond length l. From (2.20) one establishes E_h in diamond and silicon, and thereby obtains the relation

$$E_h \propto l^{-2.5}. \tag{2.21}$$

The power-law form of (2.21) of course is not unique, but generally speaking power laws seem to give good results for scaling in discussing related questions such as phase transitions. We therefore employ this functional form in (2.21).

COMPLEX ENERGY GAPS AND RESONANCE

Having established the contribution of V_{covalent} to the energy gap of A–A bonds, we turn to A–B bonds. One notices from (2.19) that $E_A \neq E_B$, so that even without taking account of V_{covalent}, there is an energy gap produced by V_{ionic}. It would be a mistake, however, to attempt to estimate this effect from atomic values of E_A and E_B given by (2.19). As we remarked earlier, the formation of bonds alters energy levels too much for the atomic values of E_A and E_B to provide an accurate picture of the energy levels of the crystal.

Instead we once again have recourse to band theory, which shows that one must sum the potentials V_{covalent} and V_{ionic} over all the cells in the crystal. This brings in the crystalline structure factor, which ultimately multiplies V_{covalent} and V_{ionic} by factors which are 90° out of phase. Thus the total energy gap E_g associated with the A–B bond in the crystal has the form

$$E_g = E_h + iC, \tag{2.22}$$

where E_h represents the average energy gap produced by V_{covalent} and C represents the magnitude of the energy gap produced by V_{ionic}. We now replace E_h^2 in (2.20) by[3,4]

$$E_g^2 = E_h^2(l_{AB}) + C^2(V_A - V_B), \tag{2.23}$$

in which the functional dependence of E_h and C is explicit.

One should note that the discussion of covalent and ionic interactions so far has involved only potentials and energy gaps, not specific wave functions. The nature of the hybridized orbitals associated with bonding and antibonding states is shown in Fig. 2.3. The bonding states have lower

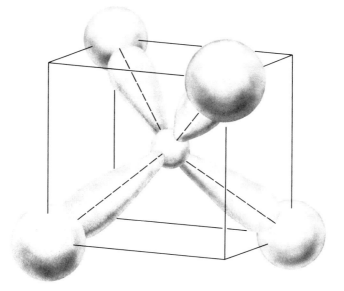

(a)

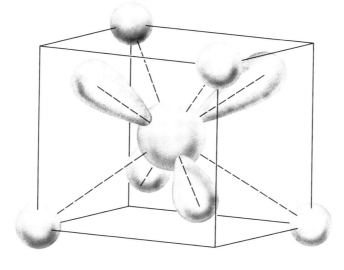

(b)

Fig. 2.3. Directed (a) bonding and (b) antibonding orbitals. The A atoms are more electropositive, the B atoms more electronegative. The bonding orbitals have lower energy both because they are directed towards their nearest-neighbors and because they are centered predominantly on the more electronegative atom. These effects are represented by the energy gaps E_h and C, respectively.

energy, are centered predominantly on the more electronegative atom (see below), and point towards the nearest-neighbor atoms. The antibonding states are centered predominantly on the more electropositive atom, and point away from the nearest-neighbors.

The relation (2.23) represents a rigorous realization of Pauling's idea that homopolar and heteropolar contributions to the partially ionic bond are resonant. The use of this terminology is justified by analogy with a simple harmonic oscillator, where kinetic and potential energies are 90° out of phase [as in (2.22)] and appear together as the sum of two squares, i.e., $p^2/2m + kx^2/2$, as in (2.23). The resonance relationship is exact only for a crystalline framework in which the structure factors for $V_{covalent}$ and V_{ionic} are 90° out of phase.

We now return to the functional dependence of E_h and C as described in equation (2.23). The bond length l_{AB}, which determines E_h, can be taken from experiment or from the radii tables of Chapter 1. However, we must still determine the dependence of C on $V_{ionic} = V_A - V_B$.

HETEROPOLAR ENERGY GAPS

The ionic contribution to the energy gap E_g is described by C, which in turn is related to the antisymmetric potential V_{ionic}. One can derive empirical values of $C(AB)$ from the observed dielectric constants of $A^N B^{8-N}$ as follows. The case $N = 4$ determines the constants in the interpolation formula (2.21) for $E_h(l_{AB})$. Replacing E_h^2 in (2.20) by E_g^2 as given by (2.23) fixes $C^2(AB)$ in terms of $\epsilon(0) = n^2(AB)$ and l_{AB}.

There are a great many tetrahedrally coordinated $A^N B^{8-N}$ crystals (about forty altogether), so that one can survey the empirical values of C_{AB} and relate them to the atomic potentials V_A and V_B. We saw in (2.1) and (2.2) that the potentials of the ion cores are Coulombic outside the core radii. To get the total potential including that of the other valence electrons, we allow the latter to screen the ion-core potential. According to the Thomas-Fermi theory, this multiplies each potential by a screening factor which is $\exp(-k_s r)$. The magnitude of k_s is discussed later, but for the moment we need only know that it is proportional to $l^{-1/2}$, where l is the bond length.

Next we must average these potentials over the region of space appropriate to bond formation. In Chapter 1 we discussed the covalent atomic radius r_A. Because l_{AB}, the A–B bond length, is given by $r_A + r_B$, it seems reasonable to evaluate V_A and V_B at r_A and r_B, respectively. Moreover, C depends on $V_A - V_B$. This suggests that we try the relation

$$C(A, B) = b(Z_A e^2/r_A - Z_B e^2/r_B) \exp(-k_s l_{AB}/2) \qquad (2.24)$$

where b is a dimensionless constant of order unity. Empirically (2.24) is found to be quite accurate [3], with the value of b equal to 1.5 with an r.m.s. uncertainty of 0.1.

One can compare (2.24) to Pauling's electronegativity differences $\Delta X_{AB} = X_A - X_B$. It turns out that for nontransition elements Pauling's values of X_A can be reproduced by

$$X_A \propto Z_A e^2/r_A + \text{const,} \qquad (2.25)$$

so that $C(A, B)$ is proportional to $X_A - X_B$ if the exponential screening factor is treated as a constant [4]. This factor actually varies however by about 40% from the first to fourth rows of the periodic table. Another way of comparing the ionic energy gaps $C(A, B)$ to Pauling's electronegativity differences ΔX_{AB} is to take the empirical values of $C(A, B)$ as determined directly from $\epsilon_0(AB)$ and plot these against ΔX_{AB}. This is done in Fig. 2.4, where the linear relation

$$C(A, B) = 5.75\Delta X_{AB}(\text{eV}) \qquad (2.26)$$

is seen to be roughly valid. Almost all the deviations from a simple linear relation can be shown to arise from neglect of the exponential screening factor in (2.24) in Pauling's definition of X_A in (2.25). For nontransition

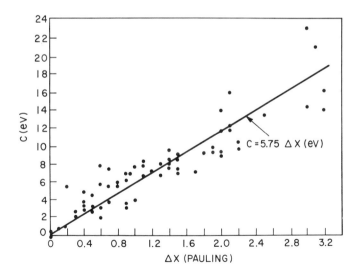

Fig. 2.4. The dielectrically determined ionic energies $C(A,B)$ against Pauling's electronegativity difference ΔX_{AB} for $A^N B^{8-N}$ crystals with fourfold or sixfold coordination.

elements of well-defined valence we conclude that $C(A, B)$ is a more meaningful measure than ΔX_{AB} of the ionic energy in tetrahedral bonds.

MODERN DEFINITION OF IONICITY

The fractional ionic character of a covalent bond f_i should depend on the magnitude of the ionic or antisymmetric potential V_{ionic} compared to the symmetric potential $V_{covalent}$. In terms of the *ionicity phase angle* φ of $E_h + iC$ shown in the (E_h, C) plane in Fig. 2.5,

$$\tan \varphi = C/E_h. \qquad (2.27)$$

This suggests the definitions

$$f_i = \sin^2 \varphi = C^2/E_g^2, \qquad (2.28)$$

$$f_h = \cos^2 \varphi = E_h^2/E_g^2 \qquad (2.29)$$

for the ionic and covalent characters of an A–B bond.

Some examples of (2.28) and (2.29) are $f_i(\text{Ge}) = 0.00$, $f_i(\text{GaAs}) = 0.31$, $f_i(\text{ZnSe}) = 0.68$. The situation is illustrated in Fig. 2.6 for Ge and ZnSe. For III–V semiconductors C is nearly always less than E_h and the bond is predominantly covalent. For most II–VI and I–VII crystals, however, C is larger than E_h and the bonding is predominantly ionic. A complete list of ionicities of $A^N B^{8-N}$ crystals in diamond, zincblende and

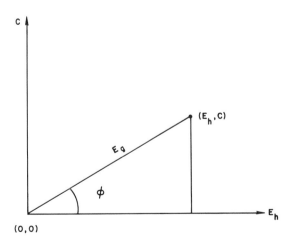

Fig. 2.5. The phase angle φ in the (E_h, C) plane measures the ionic and covalent character of $A^N B^{8-N}$ bonds.

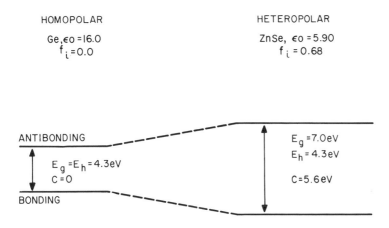

HOMOPOLAR

Ge,ϵ_0 =16.0
f_i =0.0

HETEROPOLAR

ZnSe, ϵ_0 = 5.90
f_i = 0.68

ANTIBONDING

E_g =E_h = 4.3eV
C = 0

BONDING

E_g =7.0eV
E_h = 4.3eV

C=5.6eV

Fig. 2.6. Representative values of E_h, C, and $E_g{}^2 = E_h{}^2 + C^2$ for Ge and ZnSe. Also given are the fractional ionic characters f_i.

wurtzite structures is given in Table 2.2 [5]. The statistical distribution of structures of 68 crystals of these types and of the rocksalt type in the (E_h, C) plane is shown in Fig. 2.7.

STATISTICAL TEST OF DEFINITIONS OF IONICITY

One of the problems inherent in defining ionicity of the covalent bond is the difficulty in transforming a qualitative or verbal concept into a quantitative, mathematical formula. Every definition involves some assumptions, and who is to say which assumptions are more valid than others?

Fortunately in the case of bonds in crystals of the family $A^N B^{8-N}$ there is a very simple resolution to this problem. The more ionic members of this family nearly all crystallize in the rocksalt (NaCl) structure. These crystals have been known to be quite ionic since the work of Madelung and Born around 1910. One can describe the cohesive energies, lattice constants, and bulk moduli of these crystals by a classical model based on closed-shell (rare gas) electronic configurations, e.g., Na^+Cl^-. Indeed the cohesive energy arises mainly from the Coulombic attraction of cation and anion, this force being balanced at the equilibrium lattice constant by closed-shell repulsion (exclusion principle). The charge transfer is nearly complete, and there is little electron sharing between cation and anion, as occurs in covalent crystals. Most of the electron sharing, such as it is, takes place not between cation and anion but between anion and anion.

Tetrahedrally coordinated structures are favorable to formation of sp^3

TABLE 2.2

Average Energy Gaps in Binary Tetrahedrally Coordinated Crystals

Crystal	E_h (eV)	C (eV)	E_g (eV)	f_i
C	13.5	0	13.5	0
Si	4.77	0	4.77	0
Ge	4.31	0	4.31	0
Sn	3.06	0	3.06	0
BAs	6.55	0.38	6.56	0.002
BP	7.44	0.68	7.47	0.006
BeTe	4.54	2.05	4.98	0.169
SiC	8.27	3.85	9.12	0.177
AlSb	3.53	2.07	4.14	0.250
BN	13.1	7.71	15.2	0.256
GaSb	3.55	2.10	4.12	0.261
BeSe	5.65	3.36	6.57	0.261
AlAs	4.38	2.67	5.14	0.274
BeS	6.31	3.99	7.47	0.286
AlP	4.72	3.14	5.67	0.307
GaAs	4.32	2.90	5.20	0.310
InSb	3.08	2.10	3.73	0.321
GaP	4.73	3.30	5.75	0.327
InAs	3.67	2.74	4.58	0.357
InP	3.93	3.34	5.16	0.421
AlN	8.17	7.30	11.0	0.449
GaN	7.64	7.64	10.8	0.500
MgTe	3.20	3.58	4.80	0.554
InN	5.93	6.78	8.99	0.578
BeO	11.5	13.9	18.0	0.602
ZnTe	3.59	4.48	5.74	0.609
ZnO	7.33	9.30	11.8	0.616
ZnS	4.82	6.20	7.85	0.623
ZnSe	4.29	5.60	7.05	0.630
HgTe	2.92	4.0	5.0	0.65
HgSe	3.43	5.0	6.1	0.68
CdS	3.97	5.90	7.11	0.685
CuI	3.66	5.50	6.61	0.692
CdSe	3.61	5.50	6.58	0.699
CdTe	3.08	4.90	5.79	0.717
CuBr	4.14	6.90	8.05	0.735
CuCl	4.83	8.30	9.60	0.746
CuF	8.73	15.8	18.1	0.766
AgI	3.09	5.70	6.48	0.770
MgS	3.71	7.10	8.01	0.786
MgSe	3.31	6.41	7.22	0.790
HgS	3.76	7.3	8.3	0.79

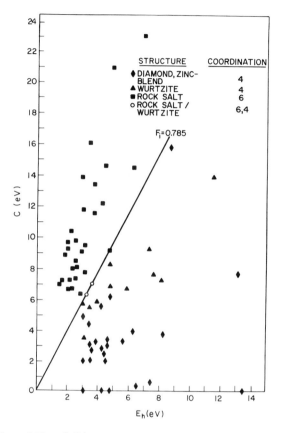

Fig. 2.7. Values of E_h and C for crystals of the type $A^N B^{8-N}$. The coordination numbers (fourfold or sixfold) are also indicated. Note that the line $F_i = 0.785$ separates all fourfold from all sixfold coordinated crystals.

bonds, but the NaCl structure, in which each ion is sixfold coordinated, gives a larger cohesive energy for ionic binding, because the Coulomb attractive energy is greater for sixfold coordination. Thus it becomes apparent that when the bond is sufficiently ionic, i.e., when $f_i(AB) > F_i$, the crystal will have a structure in which each ion is more than fourfold coordinated. Here F_i is a critical or threshold ionicity for transition from a predominantly covalent zincblende or wurtzite structure to the ionic rocksalt structure.

Any given definition of ionicity is likely to be imperfect. However, if a prescription for calculating $f_i(AB)$ is given which is independent of the coordination number of each atom in the crystal, then this prescription can be tested by regarding F_i as an independent variable. One then calculates for the approximately 70 crystals of the $A^N B^{8-N}$ family all the ionicities, arranging them in order from smallest to largest. Each value of F_i will be associated with a certain number of incorrect predictions $M(F_i)$ of coordination numbers four or six. For example, if F_i is too small many fourfold coordinated crystals would be predicted to have sixfold coordination, whereas if F_i is too large the converse would happen. The difficulties involved in predicting whether a given crystal is fourfold or sixfold coordinated are apparent from the tableau shown in Fig. 2.8, which shows the coordination numbers for $N = 1$ and 2. (All $N = 3$ and $N = 4$ crystals are fourfold coordinated).

By inspection of Fig. 2.8, we can see that by guessing that all alkali (group IA) and alkaline earth (group IIA) crystals are sixfold coordinated, while all others are fourfold coordinated, one makes nine errors. A good definition of ionicity must do significantly better than this.

The Coulson method, which is based on hybridized sp³ orbitals, is confined to crystals with fourfold coordination, and thus cannot be used to predict coordination configurations. Pauling's definition of single-bond ionicities, Eq. (2.16), is independent of coordination number. The resonating-bond definition Eq. (2.18) depends on the coordination number M. For $N = 2$, it assigns a value of f_i' to an $A^N B^{8-N}$ compound with $M = 6$ which is shifted to values larger than the value for the same compound with $M = 4$ by about 0.12. The minimum number of errors in predicting coordination number using the single-bond ionicities f_i is eight, as shown in Fig. 2.9. Neglecting the controversial Ag halides and ignoring the three crystals (MgS, MgSe, HgS) which are found with both structures, one finds a minimum number of errors of four, using f_i'. Most of the improvement in f_i' compared to f_i results from the post hoc 0.12 shift of the latter compared to the former, for $M = 6$ compared to $M = 4$.

The modern definition of ionicity Eq. (2.28) has also been used to calculate [5] $f_i(AB)$ for these crystals. It is found that with $F_i = 0.785$

	IA				IB			IIA			IIB			
	Li	Na	K	Rb	Cu	Ag		Ca	Sr	Ba	Be	Mg	Zn	Cd	Hg
F	6	6	6	6	4	6	O	6	6	6	4	6	4	6	6
Cl	6	6	6	6	4	6	S	6	6	6	4	6-4	4	4	6-4
Br	6	6	6	6	4	6	Se	6	6	6	4	6-4	4	4	4
I	6	6	6	6	4	4	Te	6	6	6	4	4	4	4	4

Fig. 2.8. Coordination numbers of $A^N B^{8-N}$ crystals for $N = 1, 2$. From the tableau one sees that CuF and MgTe are fourfold coordinated, while AgBr and MgO are sixfold coordinated.

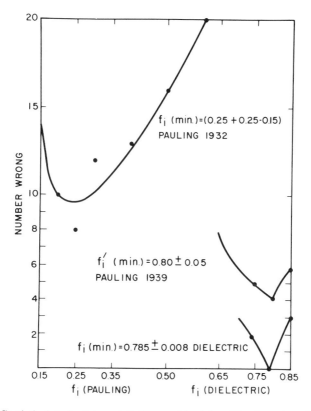

Fig. 2.9. Statistical test of two definitions of ionicity. The fact that the modern dielectric definition makes no errors with $F_i = 0.785$ is equivalent to the statement that this line divides the (E_h, C) plane into fourfold and sixfold coordinated domains in Fig. 2.7.

no errors in predicting coordination number are made. As can be seen in Fig. 2.9 this implies a statistical accuracy for the modern definition of at least 1%; i.e., the modern scale of ionicity is much more accurate than Pauling's scales and requires no input information concerning coordination numbers. It is so accurate, in fact, that it can be used to calibrate all other scales. In this way one finds [5] that the accuracy of the Coulson scale is about the same as that of Pauling's scale.

BORDERLINE CRYSTALS

Some of the most interesting examples among the 68 crystals studied are those with ionicities just below or just above the critical value. The ion-

icities of five crystals of each type are listed in Table 2.3. Note that MgS
and MgSe, which have ionicities closest to the critical value, are metastable
in both coordination configurations, while AgI, with ionicity just below the
critical value, can be transformed to the NaCl structure with a pressure of
only a few thousand atmospheres. In Chapter 8, and particularly Fig. 8.5,
such transformations are discussed in detail.

Borderline binary compounds have been found to exhibit quite remarkable structural behavior when alloyed with a suitably chosen third element.
For AgI a small admixture of an alkali metal such as Rb leads to the
formation of a ternary compound with an extraordinary crystal structure.
The ternary compound has the chemical formula $RbAg_4I_5$, and it contains
four formula units (altogether 40 atoms) per unit cell. In each unit cell
there are four groups RbI_6, i.e., four Rb ions octahedrally coordinated with
six iodine ions each, just as one would have in RbI. Each of the Ag ions is
tetrahedrally coordinated with I ions, just as they would be in AgI. However, there are 72 sites available for the Ag ions in a unit cell containing
only 16 Ag ions, so most of the Ag sites are vacant. They are also adjacent
to one another, and this leads to remarkably high values for ionic conductivity by Ag ions. Other borderline compounds, such as MgS and MgSe,
may form such "compromise" structures, e.g., when alloyed with Ba. In
each case proximity of the binary parent to the borderline between two
coordination configurations is the condition requisite to formation of the
scaffolded structure.

Although the transition from fourfold to sixfold coordination is necessarily abrupt, one would expect that the transition from electron sharing
(covalency) to electron transfer to form closed shells (classical ionic model)

TABLE 2.3

Ionic Characters of Crystals near the Borderline between Fourfold and Sixfold Coordinated
Structures

Fourfold coordinated		Sixfold coordinated	
Crystal	f_i	Crystal	f_i
CdSe	0.699	MgS[a]	0.786
CdTe	0.717	MgSe[a]	0.790
CuBr	0.735	MgO	0.841
CuCl	0.770	AgBr	0.850
AgI	0.770	AgCl	0.856

[a] Metastable in fourfold coordinated structures.

should be more diffuse. Moreover, this transition need not be centered on $F_i = 0.785$, but could be centered either below this value in the covalent structures or above it in the rocksalt ones. Evidence on this question comes from the empirical values of b in Eq. (2.24). As mentioned previously for fourfold coordination the values of b are 1.5 ± 0.1, but for ionicities greater than 90%, the average value [3] of b is about 2.25, i.e., larger than the covalent value by about 6/4, the ratio of coordination numbers. The diffuse transition region in which b changes from 1.5 to 2.25 is from about 85% to 90% ionicity. In this region there is still appreciable electron sharing and covalency, even though the crystal structures are all of the rocksalt type. In turn this means that the covalent model is valid for *all* tetrahedrally coordinated crystals, because closed shell effects are not of great importance until $f_i \gtrsim 0.85$.

It has been speculated at various times that increased ionicity, because it is associated with a tendency towards closed shell formation, might lead to an expansion of bond lengths and lattice constants. Such an expansive effect does occur, and it correlates well with the increase in b values [3]. Again this means that the effect can be neglected entirely for $f_i \lesssim 0.85$, i.e., in all tetrahedrally coordinated semiconductors described in Chapter 1 in terms of covalent radii.

TRUE (UNDISTORTED) SCALES

From Fig. 2.9 it can be seen that Pauling's single-bond value of F_i is 0.25, while the modern spectroscopic value is 0.785. The value of F_i obtained with resonating bonds is about 0.80, according to Fig. 2.9, in good agreement with the spectroscopic value. This shows that the concept of resonating bonds is a valid one, although it gives results for f_i which are less accurate and more ambiguous than the spectroscopic values. Presumably the modern definition, Eq. (2.28), which treats the symmetric and anti-symmetric or covalent and ionic effects on an equal footing in terms of the ionicity phase angle φ, and which does not introduce any arbitrary constants into the definition of either term, gives the true value for $F_i = f_i(\text{min.})$. In fact the definition (2.28) ought to define an undistorted or true scale of ionicity.

What do we mean by a true scale? An analogy may be helpful. Just as ionicity has long been considered a qualitative concept not susceptible to quantitative definition, so in the eighteenth century was heat regarded. (Indeed one cannot specify the heat content of a body.) Then it was discovered that heat exchange in a reversible process can be related to changes in entropy content, which *are* well-defined, by means of an integrating factor which is the absolute or Kelvin temperature. Thus the latter is a true temperature scale, as contrasted for instance, with the Fahrenheit scale, which is based on the thermal properties of a particular real material.

COHESIVE ENERGIES

If the definition (2.28) constitutes a true scale of ionicity, and if the latter is the dominant factor in determining the physical properties of tetrahedrally coordinated semiconductors, then these properties should exhibit a simple functional dependence on f_i, *providing* the latter is evaluated according to (2.28). (On the other hand, use of other definitions might obscure this dependence). Other factors may also play a role, and a given property may depend on E_h and C in a more complicated way than through the ionicity phase angle φ which defines $f_i = \sin^2 \varphi$. But in many cases a suitably normalized bulk property should exhibit a simple dependence on f_i.

Perhaps the most fundamental bulk property is the cohesive energy of the A–B crystal measured relative to separated A and B neutral atoms. (As remarked previously, the heat of formation Δ_{AB} at STP is a much less meaningful quantity, because at STP the constituents AA and BB may be either solid, liquid, molecular, or monatomic). This cohesive energy is denoted by ΔG_s, the Gibbs free energy of sublimation into neutral atoms at STP. There is a very extensive literature of values of ΔG_s for semiconductors, bibliographies of which are available from the U.S. National Bureau of Standards. Recent values, which are quoted below, generally agree with one another to within about 1%, which is also the level of accuracy of the values of f_i determined from experimental data.

In general we expect that $\Delta G_s(AB)$ will be a function of $f_i(AB)$ and of $l(AB)$ which determines the density of the crystal. When A^N and B^{8-N} both belong to a given row of the periodic table (e.g., the third row, which contains Ge, GaAs and ZnSe), then $l(AB) = l(N)$ is nearly constant for $N = 2, 3, 4$ (see Table 1.5). Such a sequence of crystals can be called a horizontal sequence. One can also form quasi-horizontal sequences where one atom belongs, e.g., to row 2 and the other to row 3. Thus both GaP and AlAs belong to a 2–3 quasi-horizontal sequence.

The cohesive energies of row 1, 2, 3, and 4 horizontal sequences and quasi-horizontal sequences not involving row 1 are all shown in Fig. 2.10. All the cohesive energies are [5] very nearly linear in f_i. A particularly striking case is the row 1 horizontal sequence diamond–BN–BeO, where a straight line through the first two crystals predicts ΔG_s in BeO to within 1.5%, i.e., to within experimental uncertainty. Similar results are found for

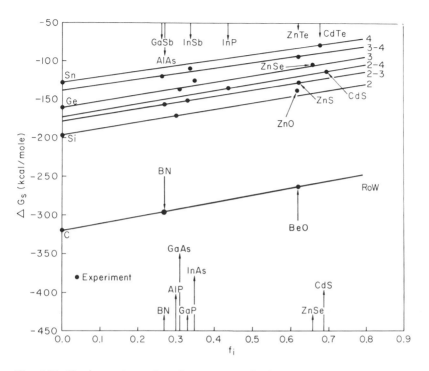

Fig. 2.10. To demonstrate that the spectroscopic theory of ionicity defines a *true ionicity scale*, the Gibbs free energies of atomization ΔG_s at STP are plotted against f_i defined by Eq. (2.28).

other sequences, all of which follow the relation

$$\Delta G_s(f_i) = \Delta G_s(0) + k f_i. \qquad (2.30)$$

An interesting feature of Fig. 2.10 is that in a given horizontal or quasi-horizontal sequence the cohesive energy decreases as f_i increases, i.e., *in predominantly covalent tetrahedral structures the binding energy is greatest for zero ionicity*. It is interesting to discuss various qualitative explanations for this trend.

The classical valence explanation, used by Pauling, is the following. As one passes from C_2 to BN to BeO, i.e., from $N = 4$ to $N = 2$ in a horizontal sequence $A^N B^{8-N}$, the number of valence electrons free to participate in the covalent bond is reduced from four to two, and there is a corresponding reduction in the number of valence bonds per atom pair. However, according to Eq. (2.15), with increasing ionicity the heat of formation per bond increases. The latter compensates partially for the former, but there is still a decrease in cohesive energy between diamond and BeO. In order to account for the tetrahedral coordination of BeO, Pauling assumes that although there are only two bonds present at any one time, these "resonate" from one position to another, effectively replacing the four bonds of the diamond lattice in a time-averaged sense.

Slater points out that there are two possible ideal configurations, $Be^{2+}O^{2-}$ and $Be^{2-}O^{2+}$, the former corresponding to closed shells and being fully ionic, the latter corresponding to four valence electrons on each atom and being fully covalent. The actual configuration in the crystal is expected to be a compromise between the two possibilities, each atom being approximately neutral.

The classical or resonating bond picture of Pauling is not meant to be taken literally. Optical data indicate that the energy spectrum of the valence electrons in BeO does not separate into two parts of equal weight, one of which corresponds to the diamond spectrum and the other of which corresponds to electrons localized about the O ion. Instead, because of the presence of V_{ionic}, the spectrum as a whole shifts to higher energies, because the average energy gap between bonding and antibonding states has increased as described by (2.23). The increase in magnitude of C compared to E_h does lead to greater probability of an electron in a bonding state being centered on the anion sublattice rather than the cation sublattice, the relative probabilities being $1 + (C/E_g)^2$ compared to $1 - (C/E_g)^2$. All of these quantities are defined in a spectroscopic way, however, which is much more accurate than the thermochemical way. Indeed the fractional ionicity and covalency defined by (2.28) and (2.29) provide spectroscopic calipers for the compromise described by Slater.

The accuracy of the classical valence bond model for an $A^N B^{8-N}$ crystal can also be examined statistically. Pauling's equation (2.15) for the contribution of extraionic energy to the heat of formation may be written

$$\Delta H(A^N B^{8-N}) = -23(\text{kcal/mole}) M(X_A - X_B)^2 + 55.4 n_N + 26.0 n_O,$$

$$(2.31)$$

where M is the number of resonating bonds per AB pair, and n_N and n_O are the number of N and O ions in the AB pair. One can solve (2.31) to obtain empirical values of M for the case $N = 2$ and for the case $N = 3$. According to the resonating bond theory $M = N$. One finds in this way [5] $N = 2$, $\bar{M} = 2.2 \pm 1.2$, $N = 3$, $\bar{M} = 4.2 \pm 1.5$. At least for the more ionic II–VI crystals $M = N$ is good, but for the less ionic III–V crystals better results come from assuming $M = 4$, corresponding to four single bonds as in diamond or Si. The scatter, as measured by the quoted rms errors, is rather large in both cases.

ITINERANT CHARACTER OF COVALENT BINDING

If one makes the Hartree or one-electron approximation, then the complete wave function Ψ which describes the motion of all the electrons in the crystal can be written as a product of one electron wave functions $\psi_i(\mathbf{r}_i)$:

$$\Psi = \prod_i \psi_i(\mathbf{r}_i). \tag{2.32}$$

The Hartree approximation, or other very similar approximations of a one-particle type, is absolutely necessary for practical calculations based on Schrödinger's equation. In certain circumstances it also gives a good description of the behavior of semiconductors in the presence of either static or optical frequency electromagnetic fields. However, this by no means implies that the single-particle approach based on (2.32) will give good estimates of crystalline binding energies. In fact no one has yet succeeded in carrying through such calculations for a range of covalent crystals.

The remarkable regularities exhibited in Fig. 2.10 can be understood if we can explain: (1) the dependence of $\Delta G_s(f_i = 0)$ on row number R, and (2) the slope $k(R)$.

From the point of view of "universal" semiconductors, the difference between one row and the next is primarily a difference in lattice constant or bond length $l_{44}(R)$. We therefore determine the dependences of $\Delta G_s(0)$ and k on l, assuming power-law scaling:

$$\Delta G_s(0) \propto l^{-n_1}, \qquad k \propto l^{-n_2}. \tag{2.33}$$

The best value for the scaling index n_1 is 2, i.e., $\Delta G_s(0)$ scales like a kinetic energy. A convenient unit of kinetic energy is E_F, the Fermi energy of a free electron gas of density equal to that of the valence electrons. The ratio $\Delta G_s(0)/E_F$ is listed in Table 2.4. If the power law (2.33) were strictly valid, this ratio would be constant. The deviations from constancy arise from core effects discussed below.

The scaling index n_2 is much smaller and to a first approximation it appears that $n_2 = 0$. However, $k(1) = 100$ kcal/mole and $k(R)$ for $R \geq 2$ is about 80–85 kcal/mole. This corresponds to $n_2 = 0.4$ or 0.5.

These two scaling results suggest the following picture for the binding energy of tetrahedrally coordinated crystals. For the homopolar values $\Delta G_s(0)$ it appears that most of the binding energy has kinetic character and is therefore associated with delocalized (itinerant) valence electrons. This confirms the relevance of the single particle approximation (2.32), because even though the one-electron functions ψ_i have not been used to calculate ΔG, the translational periodicity of the crystal requires ψ_i to

TABLE 2.4

Comparison of Cohesive Energies in Diamond-
Type Crystals with Fermi Energy of Free-
Electron Gas of Density Equal to That of
the Valence Electrons

	E_F (kcal/mole)	ΔG_s (kcal/mole)	$\Delta G_s/E_F$
C	667	320	0.478
Si	287	197	0.685
Ge	265	161	0.607
Sn	210	128	0.608

have the Bloch or delocalized form (Chapter 6). If similar scaling arguments are applied to small molecules (such as MH_3, where $M = N$, P, As
or Sb) one finds $n_1 = 1$; i.e., the binding energy scales like a potential
energy, as one would expect for localized bonds. Thus scaling provides a
simple way of showing that localized bonds are appropriate for discussing
cohesive energies in molecules, but that the valence electrons in covalent
crystals are itinerant, in much the same way as in metals like Na or Al.

The second scaling result concerns $k(l)$, the slope which describes the
reduction in cohesive energy with increasing magnitude of V_{ionic} compared
to $V_{covalent}$, i.e., C compared to E_h. The fact that $n_2 \approx 0.5$ suggests that
$| \Delta G |$ is reduced because energy is required to screen V_{ionic}. Dielectric
screening can be described at short wave lengths by the Thomas-Fermi
screening wave length $\lambda = k_s^{-1}$ [see also equation (2.24)]. This suggests

$$k(l) \propto V_{ionic}(l/\lambda) \qquad (2.34)$$

since when λ is large, little screening of V_{ionic} has taken place. Of course,
$V_{ionic} \propto l^{-1}$, and the Thomas–Fermi formula shows $\lambda \propto l^{-1/2}$. Thus (2.34)
gives $n_2 = 0.5$, as observed experimentally.

CORE CORRECTIONS

The "universal" semiconductor model we have been discussing neglects
differences between atomic cores insofar as possible. In the last chapter
these differences were used to construct rationalized tetrahedral radii, but
these are already incorporated into f_i though E_h and through C. The
remaining core effects are small, but still explicable.

We saw in Table 2.4 that $l^2 G_s(0)$ is nearly constant, but that it reaches a peak for Si. The reason for this is well known. In diamond bonding electrons are almost confined to the subspace of 2s and 2p atomic orbitals, because the promotion energy to 3d and 4f states is large. But in Si the 3p–3d promotion energy is much smaller, and this extra degree of freedom increases ΔG_s in Si over what it was in sp^3–hybridized diamond. On the other hand, in Ge and Sn the valence electrons are squeezed between d core states on each atom (these were absent in Si). Because of the exclusion principle this pushes the valence d energy up, increasing the p–d promotion energy. Thus $l^2 \Delta G_s$ is 50% larger in Si than in diamond, but only 35% larger in Ge and Sn than in diamond.

Core effects also partially affect the linear relation between ΔG_s and f_i when the cores of atoms A and B are not isoelectronic, i.e., when the sequence is not a horizontal one made up of atoms from the same row of the periodic table. One of the worst cases shown in Fig. 2.10 is the 3–4 row. Here GaSb falls above the line by six kcal/mole, and InAs falls below it by the same amount. Note that in both cases ΔG_s falls closer to the linear relation of the horizontal sequence of the anion (the row 4 sequence for GaSb, the row 3 sequence for InAs). This is a localization effect of the sort treated by localized bond models, but it is quite small compared to a typical value for $k(R)$, which is 80–100 kcal/mole. After allowance is made for this kind of core correction, $\Delta G_s(R_1, R_2)$ is indeed linear in f_i to about 1%.

ELECTRONEGATIVITY TABLE

Pauling's table of elemental electronegativities is widely used, and is frequently valuable for qualitative considerations. The most serious criticism of the table is that one would expect that the "ability of an atom to attract electrons to itself" varies, depending on whether the electron is in an s–p state, or d or f states. This variation is described as the orbital dependence of electronegativity. It introduces considerable uncertainties in discussing compounds containing transition and nontransition elements.

These uncertainties are greatly reduced if we confine ourselves to tetrahedrally coordinated compounds in sp^3-hybridized states. Then it is worthwhile to prepare a table similar to Pauling's, but which includes the effect of valence electron screening. As we saw in connection with Fig. 2.4, with Pauling's definition of X_A,

$$X_A{}^N - X_B{}^{8-N} \propto (N/r_A) - (8 - N)/r_B, \qquad (2.35)$$

whereas our more accurate equation for C_{AB}, (2.24), includes valence screening through the Thomas-Fermi factor $\exp(-k_s R)$, with $R = (r_A + r_B)/2$. For qualitative purposes we can make the approximation

$$C_{AB} \propto \{(N/r_A) - [(8 - N)/r_B]\} \exp(-k_s R)$$

$$\rightarrow \{(N/r_A) \exp(-k_s r_A)$$

$$- [(8 - N)/r_B] \exp(-k_s r_B)\} \propto X_A - X_B, \quad (2.36)$$

so that our definition of an elemental electronegativity X_A differs from Pauling's through the factor $\exp(-k_s r_A)$. One can then adjust the scale so that the two definitions agree for elements in the Li–F row of the periodic table. The differences between the two definitions then appear in the values for other rows of the periodic table.

These values are shown in Table 2.5, with Pauling's values shown in parentheses. Because $k_s r_A \propto r_A^{1/2}$, the dominant factor in our definition of X_A^N is the same as the factor used by Pauling, i.e., N/r_A. However, the differences between our values and Pauling's are significant in some cases, e.g., in comparing atoms from the first row with those from other rows. Thus when Be impurities are introduced substitutionally, replacing Ga in GaP, the dielectric scale gives $X_{Ga} < X_{Be} < X_P$, whereas Pauling's scale gives $X_{Be} < X_{Ga} < X_P$. The dipole moment associated with the vibration

TABLE 2.5

Elemental Electronegativities of Nontransition Atoms in Tetrahedrally Coordinated Environments[a]

Li	Be	B	C	N	O	F
1.00	1.50	2.00	2.50	3.00	3.50	4.00
Na	Mg	Al	Si	P	S	Cl
0.72	0.95	1.18	1.41	1.64	1.87	2.10
(0.9)	(1.2)	(1.5)	(1.8)	(2.1)	(2.5)	(3.0)
Cu	Zn	Ga	Ge	As	Se	Br
0.79	0.91	1.13	1.35	1.57	1.79	2.01
(1.9)	(1.6)	(1.6)	(1.8)	(2.0)	(2.4)	(2.8)
Ag	Cd	In	Sn	Sb	Te	I
0.57	0.83	0.99	1.15	1.31	1.47	1.63
(1.9)	(1.7)	(1.7)	(1.8)	(1.9)	(2.1)	(2.5)
Au	Hg	Tl	Pb	Bi		
0.64	0.79	0.94	1.09	1.24		
(2.4)	(1.9)	(1.8)	(1.8)	(1.9)		

[a] For the first row these have been scaled to agree with Pauling's definition. For other rows Pauling's values are also given (in parentheses).

of the Be atom is expected to be small in the former case, because the charge flow from Be to Ga is opposite in sign to that from Be to P, but large in the latter case, where the two charge flow effects add. Experimentally the dipole moment is very small. Other examples will appear in specific compounds discussed in later chapters.

HISTORICAL NOTE

The explanation of the transformation from covalent to ionic structures described here is different from the one given by classical crystallographers and chemists. The classical explanation regards each atom as a rigid sphere with a radius determined by whether the crystal is "ionic" or "covalent." This spheroidal model was proposed by Hooke and by Huygens in the latter part of the seventeenth century, and was criticized for "begging the question" by Newton in his *Optics* (1704). Nevertheless, the rigid ion or closed shell model has proved useful in describing the structures of many naturally occurring minerals, almost all of which are probably more than 90% ionic. Today our interests are focused on the much less ionic semiconductors. Most semiconductors occur only rarely in nature as large crystals, but they can be grown in the laboratory. From the viewpoint of semiconductors the transformation from covalent to ionic binding is a gradual one, and many aspects of its continuous character are discussed in this book. The reader who is interested in the development of rigid-ion or billiard-ball models of crystal structure (compared to polarizability models) will find a lengthier account elsewhere [6].

SUMMARY

The connection between covalent bonds and atomic energy levels and potentials has been described in terms of hybridized orbitals for $A^N B^{8-N}$ semiconductors and insulators. Ionicity is shown to be the factor which accounts for the trend from semiconducting to insulating behavior in this structural family. It is also the factor responsible for the phase transition from fourfold to sixfold coordinated structures. Quantitative values for the strength of covalent and ionic effects are obtained in terms of E_h, the homopolar energy gap, and C, the charge transfer or ionic energy gap. The average gap between the valence and conduction bands is $E_g = |\, E_h + iC\,|$.

The cohesive energies of tetrahedrally coordinated crystals decrease with increasing ionicity. This localizes valence electrons near the anions and reduces their itinerant character, which is the chief source of cohesion in fourfold coordinated crystals. A number of other properties of these crystals depend more strongly on ionicity than on any other single factor. Many of these properties are dealt with in succeeding chapters.

REFERENCES

1. C. A. Coulson, L. R. Redei and D. Stocker, *Proc. Roy. Soc.* **270,** 357 (1962).
2. L. Pauling, "The Nature of the Chemical Bond," p. 91 ff. Cornell Univ. Press, Ithaca, New York, 1960.
3. J. A. Van Vechten, *Phys. Rev.* **182,** 891 (1969).
4. J. C. Phillips, *Phys. Rev. Lett.* **20,** 550 (1968).
5. J. C. Phillips, *Rev. Mod. Phys.* **42,** 317 (1970).
6. J. C. Phillips, *Proc. Int. School Phys. "Enrico Fermi"* Course 49 (1973).

3

Elastic and Piezoelectric Constants

In the preceding chapters we have discussed the effects of covalent and ionic forces on crystal structures and cohesive energies of semiconductors. Further insight into the nature of these forces can be obtained by studying elastic constants. This study proves especially fruitful in the diamond and zincblende structures where cubic symmetry reduces the number of independent forces. Again the importance of covalent bonding becomes apparent, and quantitative values of bond strength are obtained.

STRESSES AND STRAINS

The macroscopic theory of the elastic properties of solids is described, e.g., by Kittel [1]. The elastic strain tensor e_{ij} which describes the displacement of the nth unit cell initially situated at r^n to $r^n + \delta r^n$ is given by

$$\delta r_i^n = e_{ij} r_j^n. \qquad (3.1)$$

Here e_{ij} is a symmetric tensor of the second rank, with six independent components (three diagonal, three off-diagonal). For cubic symmetry there are only two independent components, the diagonal one describing dilation and the off-diagonal one describing shear.

The stress tensor s_{ij} is also symmetric and of second rank. The index i labels the component of the applied force, the index j labels the component of the vector normal to the surface across which the force is applied. The elastic constants c_{ijkl} constitute a fourth rank tensor relating stress to strain,

$$s_{ij} = c_{ijkl}e_{kl}. \tag{3.2}$$

Because s_{ij} and e_{kl} may each contain six independent components, c_{ijkl} may contain 36 components. These can be reduced to 21 independent ones by analysis of the strain energy density [1]. For practical purposes this is still far too many independent components to permit an incisive analysis of the microscopic forces between atoms, and one must appeal to crystal symmetry to reduce the number of independent components to a manageable limit.

The greatest reduction obtains for cubic crystals. The independent components of e_{ij} are e_{xx} and e_{xy}, while those of s_{ij} are s_{xx} and s_{xy}. These are related by

$$s_{xx} = c_{11}e_{xx} + c_{12}(e_{yy} + e_{zz}), \qquad s_{xy} = c_{44}e_{xy}. \tag{3.3}$$

The bulk modulus $B = -V\,dp/dV$ for a cubic crystal is given by

$$B = (c_{11} + 2c_{12})/3. \tag{3.4}$$

In general the lattice vibrations of a crystal can be analyzed into normal modes. At low frequencies the linear relation $\omega = ck = 2\pi c/\lambda$ holds, where k is the wave number, c the velocity of sound and λ the wavelength of the lattice wave. The general case of lattice waves is discussed in the following chapter. Here we are concerned with the elastic limit, $\lambda \gg a$, where a is a lattice constant. In this limit the linear relation $\omega = ck$ holds, and atomic displacements can be described in terms of macroscopic strains. This makes possible the determination of elastic constants by measuring transit times of sound waves whose velocities depend on the direction of k and polarization of the wave. For instance, waves polarized along (001) propagating along (100) or (110) axes travel with a velocity determined by c_{44}, while the (1$\bar{1}$0) shear wave traveling along (110) involves $c_{11} - c_{12}$. Because transit times can be measured electronically as echoes with high precision, some of our most precise data on lattice forces in semiconductors are derived this way. The samples used to obtain the data must be free of structural defects, particularly vacancies.

HARMONIC STRAIN ENERGY

Suppose that the mth nucleus in the unit cell is displaced by an amount $u_a{}^m$ along the ath coordinate axis. The total energy of the crystal will change by an amount

$$V = \tfrac{1}{2}K_{aa'}^{mm'}u_a{}^m u_{a'}{}^{m'} + 0(u^3) \tag{3.5}$$

in the presence of displacements $u_a{}^m$ and $u_{a'}{}^{m'}$. Retaining only the second-order terms in (3.5) gives the harmonic approximation. The first order terms vanish of course because $u_a{}^m$ measures the displacement from equilibrium, while the third and higher order terms give rise to decay of the harmonic normal modes which is usually small and of little structural interest.

In the presence of a general set of displacements $\{u_a{}^m\}$ the harmonic strain energy is

$$V_2 = \tfrac{1}{2} \sum_{\substack{mm' \\ aa'}} K_{aa'}^{mm'}u_a{}^m u_{a'}{}^{m'}. \tag{3.6}$$

In order to relate the elastic constants to the coefficients $K_{aa'}^{mm'}$ one calculates the velocities of sound waves both macroscopically [starting from (3.3)] and microscopically [starting from (3.6)]. The microscopic calculation is facilitated [1] by noting that for elastic waves in a monatomic lattice

$$e_{xx} = \partial u_x/\partial x \tag{3.7}$$

with similar relationships for other derivatives. In diamond one has two sublattices, so that besides (3.7) one needs another relationship to specify the internal strain or relative displacement of the sublattices. One can derive this internal relationship from (3.6) by demanding that the two sublattices satisfy a condition of relative equilibrium [2].

INVARIANCE CONDITIONS

From the foregoing discussion we see that given the coefficients K_{ab}^{mn} it is merely a matter of algebra to derive the elastic constants. Nevertheless it was not until 1966 that a fully satisfactory discussion of these coefficients for the diamond lattice was given by Keating [2], although Max Born had produced a model for the diamond lattice in 1912.

Why did construction of a satisfactory model for such a simple structure take so long? The point is that the coefficients K_{ab}^{mn} must satisfy certain symmetry conditions. Firstly they must give rise to no additional energy if the lattice is shifted uniformly, $u_a{}^m = u$. This gives the condition of translational invariance,

$$\sum_{m} K_{aa'}^{mm'} = 0, \tag{3.8}$$

which is easy to check. Another type of condition is obtained by considering rotations of the crystal. If a rotation of the crystal interchanges coordinates a and c of an atom located at $\mathbf{X}^m$, then the condition of rotational invariance

$$K_{aa'}^{mm'} X_c{}^{m'} = K_{ca'}^{mm'} X_a{}^{m'} \tag{3.9}$$

must also hold. In principle the many different cases of (3.9) can be checked by enumeration, but in practice most workers, including Born, have made errors in attempting to exhaust all the applications of (3.9) to restrict the form of $K_{aa'}^{mm'}$ to obtain a physically consistent model.

Keating used another approach, which is much more practical in cases where the forces are of short range. He *constructed* (3.6) from manifestly invariant quantities, namely scalar invariants formed from the *differences* between displaced coordinates

$$\mathbf{X}_{kl} = \mathbf{X}_k - \mathbf{X}_l. \tag{3.10}$$

Because $\mathbf{X}_{kl}$ is a vector, the desired scalar invariants have the form

$$\lambda_{klmn} = \mathbf{X}_{kl} \cdot \mathbf{X}_{mn} - \mathbf{X}_{kl}^0 \cdot \mathbf{X}_{mn}^0, \tag{3.11}$$

the second term representing the equilibrium positions. In general (3.11) is first-order in u, so that V_2 is second-order in λ_{klmn}.

MODEL FORCE FIELDS

To reduce the number of parameters still further, one can make various physical assumptions. One of these is to assume only pairwise central forces, so that only terms involving λ_{klkl} are non-vanishing. This is a plausible approximation for ionic crystals, where core–core repulsive and electrostatic forces, which are of this type, dominate. For cubic crystals in which each atom is at a center of symmetry Cauchy derived the relation

$$c_{12} = c_{44} \text{ (central forces)}. \tag{3.12}$$

In crystals of the NaCl type deviations from (3.12) result from directional interatomic forces and are usually regarded as a measure of covalency. Note that (3.12) does not hold for tetrahedrally coordinated cubic crystals, because the atoms are not at centers of inversion.

For covalent crystals in any event central forces only are poor. In molecular theory it is customary to describe the vibrations of covalent molecules (e.g., hydrocarbons) using a *valence force field*. The parameters of the valence force field are bond lengths and bond angles, and these are expanded in terms of nearest-neighbors, next nearest-neighbors and so on. Bond-stretching forces correspond to (3.11) with $mn = kl$, and bond-bending forces to $mn \neq kl$.

Bond-bending forces originate, of course, from the nature of hybridized orbitals. In tetrahedral crystals, for example, the sp^3-hybrid orbitals must form tetrahedral angles of about 110° with each other. Altering these angles means mixing into the space of sp^3-valence wave functions other wave functions of d and f symmetry. The promotion energy from 2s and 2p valence states to 3d and 4f states is very large, and for this reason we expect bond-bending forces to be especially large for first-row atoms such as carbon. The promotion energy from 3s and 3p to 3d is much smaller, so that bond-bending forces should be smaller in Si or in general for atoms from rows other than the first row.

Bond-bending forces are responsible for stabilizing open covalent structures and preventing their collapse into denser structures with the eight- or twelvefold coordinations favored by metals and very ionic crystals such as CsI. In general such collapse takes place through unstable shear modes. This is most easily seen by assuming that the only interatomic force is a central one between nearest-neighbors. In a simple cubic lattice one can then show that for the shear modes in the long-wavelength limit $\omega^2 = 0$ to order k^2, whereas one must have $\omega^2 > 0$ to have stable equilibrium. A similar result is found for diamond- and zincblende-type lattices, as one can see from the explicit formulas given below for the shear elastic constants c_{44} and $c_{11} - c_{12}$.

DIAMOND LATTICE

Denote a central atom by O and its four nearest-neighbors by $l = 1, \dots, 4$ (see Fig. 3.1). The nearest-neighbor bond-stretching contribution to V_2 is proportional to $\alpha \lambda^2_{olol}$, the bond-bending contribution is proportional to $\beta \lambda^2_{olol'}$. With only these two contributions to V_2 Keating finds in terms of the cubic lattice constant a

$$c_{11} = (\alpha + 3\beta)/4a, \qquad (3.13)$$

$$c_{12} = (\alpha - \beta)/4a, \qquad (3.14)$$

$$c_{44} = \alpha\beta/a(\alpha + \beta). \qquad (3.15)$$

Central forces correspond to $\beta = 0$, but as noted previously because the atoms are tetrahedrally coordinated the Cauchy relation (3.12) does not hold.

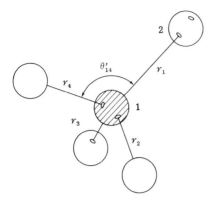

Fig. 3.1. In a tetrahedral unit the bond stretching or central forces involve changes in the bond lengths $r_1, \ldots, r_4$. The bond-bending or directional forces involve changes in $\mathbf{r}_1 \cdot \mathbf{r}_4 = r_1 r_4 \cos \theta_{14}$, as illustrated.

One can eliminate α and β from (3.13)–(3.15) to obtain a relation between c_{11}, c_{12} and c_{44} which is a check on the validity of truncating V_2 at this point. The relation is

$$2c_{44}(c_{11} + c_{12})/(c_{11} - c_{12})(c_{11} + 3c_{12}) = 1. \tag{3.16}$$

The relations (3.13)–(3.16) are checked in Table 3.1 for diamond, Si, and Ge. The agreement between theory and experiment is excellent. The ratio β/α measures the importance of bond-bending or directional forces relative to bond-stretching or central forces. This ratio is about the same in Si and Ge, but in diamond it is about twice as large. This illustrates the additional stability of tetrahedral bonds between first-row atoms arising from the large 2p-3d promotion energy.

ZINCBLENDE LATTICE

When the two atoms in the unit cell are different, as they are in crystals of the zincblende type such as BN and GaAs, a new feature enters the problem. In addition to the short-range forces described by α and β in Fig. 3.1 there are long-range Coulomb forces between the cations and the anions. The magnitude of these forces is determined by certain effective charges $\pm e^*$ which are associated with each cation or anion, respectively. The magnitude of these charges can be determined by studying the optic modes of vibration of the lattice as $k \to 0$. The way in which this is done is described in the following chapter.

The long-range Coulomb forces contribute to the elastic constants. In the zincblende lattice the expressions (3.13)–(3.15) for the diamond lattice are generalized by Martin [3] to include the combined effects of Coulomb forces and internal strain, which are coupled together through the internal electric field. Let us define an internal strain parameter by ζ (the significance of ζ is discussed below). With the abbreviations

$$s = (e^*)^2/\epsilon_0\, d^4, \qquad d = (3^{1/2})a/4, \tag{3.17}$$

where ϵ_0 is the optical dielectric constant of the crystal, one finds that

$$\zeta = (2c_{12} - 0.31s)/(c_{11} + c_{12} - 0.31s) \tag{3.18}$$

and that the elastic constants are given by

$$c_{11} = (\alpha + 3\beta)/4a - 0.08s \tag{3.19}$$

$$c_{12} = (\alpha - \beta)/4a - 0.14s \tag{3.20}$$

$$c_{44} = (\alpha + \beta)/4a - 0.14s - c\zeta^2 \tag{3.21}$$

$$c = (\alpha + \beta)/4a - 0.27s. \tag{3.22}$$

Once again α and β can be eliminated from these equations to give a relation

TABLE 3.1

Valence Forces in Diamond- and Zinc-blende-Type Crystals

Crystal	Relation[a]	β/α[b]	ζ[b]
Diamond	1.002	0.655	0.208
Si	1.005	0.285	0.557
Ge	1.081	0.294	0.546
AlSb	1.056	0.192	0.649
GaP	1.044	0.221	0.589
GaAs	1.068	0.217	0.600
GaSb	1.059	0.218	0.612
InP	1.075	0.145	0.699
InAs	1.109	0.156	0.682
InSb	1.103	0.161	0.695
ZnS	1.082	0.107	0.736
ZnSe	1.110	0.120	0.723
ZnTe	1.059	0.142	0.706
CdTe	1.045	0.084	0.793
CuCl	1.532	0.079	0.785

[a] "Relation" refers to expressions (3.16) and (3.23) for diamond and zincblende materials, respectively; the value should be 1.00 if the model holds exactly.

[b] Both β/α and ζ are discussed in the text.

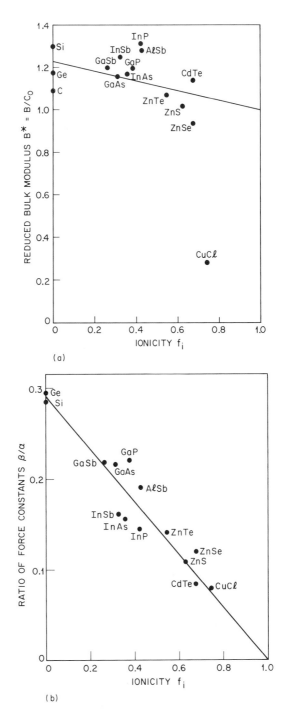

(a)

(b)

between the elastic constants which involves only the value of e^* determined from other experiments. This relation is

$$1 = \frac{2c_{44}(c_{11} + c_{12} - 0.31s)}{(c_{11} - c_{12})(c_{11} + 3c_{12} - 0.62s) + 0.26s(c_{11} + c_{12} - 0.31s)} \quad (3.23)$$

which of course reduces to (3.16) when $e^* = 0$. Martin's values of (3.23) are also listed in Table 3.1. The zincblende relation is as well satisfied as the diamond one.

SHEAR CONSTANTS AND IONICITY

Although particular shear waves are described by c_{44} and $c_{11} - c_{12}$, both of these elastic constants are proportional to β. Thus bond-bending forces stabilize tetrahedrally coordinated lattices against shear strains represented by the off-diagonal elements of the strain tensor. The ionic potential $V_A - V_B$ acts to reduce the strength of bond-bending forces by localizing more electrons around the anion and fewer around the cation. As a result the sp^3-bonding states of the cation are depopulated, while the sp^3-antibonding states of the anion are partially occupied, thereby in effect cancelling the directional character of the corresponding bonding states. Thus maximum directed bonding is achieved when the bond ionicity $f_i = 0$, so that each atom has $N = 4$ and each sp^3-bonding orbital is singly occupied, overlapping with a singly occupied orbital of its nearest-neighbor, the two electrons having zero total spin. Thus as the bond ionicity f_i increases the directional character of the bonding is reduced and β should tend to zero. The shear constants $c_{11} - c_{12}$ and c_{44} are both proportional to β, so that both shear constants are expected to decrease as f_i increases.

The bulk moduli B and the ratio β/α of bending to stretching force constants are plotted for zincblende crystals against f_i in Fig. 3.2a and Fig. 3.2b. The elastic constants are measured in units of e^2/d^4, where d is the bond length in each case. We see that B decreases slowly and irregularly as f_i increases. But as the ionicity f_i increases there is a rapid and dramatic drop in β/α. In the limit $f_i = 1.00$, we would expect this ratio to be nearly zero, as the directional forces between closed shell ions are known to be very small from studies of rare gas solids. The trend in β/α is very nearly

---

Fig. 3.2. Chemical trends in the elastic properties of zincblende crystals as a function of spectroscopic ionicity. In (a) the trend for the bulk modulus B = $(c_{11} + 2c_{12})/3$ = $(\alpha + \beta/3)4a$ is shown, while in (b) the trend is shown for β/α, which determines the shear elastic constants according to Eq. (3.13)–(3.15).

linear, and it does extrapolate to zero at $f_i = 1.00$. Thus our physical expectations are borne out, and the interpretation of f_i as a true (undistorted) scale for ionicity of the tetrahedral bond is confirmed.

An important feature of Fig. 3.2(b) which deserves comment is the *continuous* character of β/α as a function of f_i. In many chemical and crystallographic models of crystal structure and interatomic forces in $A^N B^{8-N}$ crystals, it is the *discontinuous* variable N which is of central importance. For example, in the rigid-ion or billiard-ball model of crystal structure, the transition from tetrahedral to rocksalt structures is explained in the following way. All $N = 3$ and $N = 4$ crystals are assumed to be covalent and hence exhibit tetrahedral coordination. All $N = 1$ crystals are assumed to have the NaCl or CsCl structures (six- or eight-fold coordinated), leaving compounds like CuBr and AgI (which are four-fold coordinated) as anomalies. An effort is made to explain the choices of coordination configuration only for the $N = 2$ crystals. These can be explained by assuming that when the ratio r_+/r_- of cation to anion radii is low enough, four-fold coordination is preferred. (The magic value is $r_+/r_- < 0.33$). Although this criterion works well enough for the $N = 2$ compounds, when it is applied to $N = 1$ compounds it predicts that about one-third of them should have covalent structures. Conversely, when applied to $N = 4$ compounds, it can explain the diamond structure only on the supposition that the crystal consists of alternating sublattices of C^{4+} and C^{4-} ions, an absurd hypothesis. Thus the "radius ratio" approach is meaningful only when confined to $N = 2$ compounds. However, according to Fig. 3.2, the interatomic forces are continuous functions of f_i, and there is nothing special about $N = 2$ compounds to differentiate them discontinuously from $N = 1$ and $N = 3$ compounds.

The crystal which is closest to the critical ionicity, CuCl, exhibits anomalously low elastic constants, although β/α behaves well. It is possible that we have here evidence for the instability responsible for the first-order phase transition to the rocksalt structure.

INTERNAL STRAINS

In the diamond or zincblende lattices the positions of all the atoms are uniquely determined by the cubic lattice constant. If the crystal is subjected to a uniaxial stress because there are two atoms in the unit cell this is no longer the case in general [4]. The displacement of the kth atom in the nth unit cell is given by

$$\delta r_i^{nk} = \delta r_i^n + u_i^k, \qquad (3.24)$$

where δr_i^n is given by (3.1). The internal strain displacement u_i^k is given by

$$u_i^k = \gamma_{imn}^k e_{mn}, \tag{3.25}$$

where γ_{imn}^k is a third-rank tensor whose independent components are restricted by the symmetry of the kth lattice site [5].

For diamond and zincblende crystals internal strains are easily visualized in several cases. Suppose the strain is applied normal to a (100) lattice plane. Then it makes equal angles with the $\langle 111 \rangle$ tetrahedral bonds, and each of them is equally compressed. Thus a (100) strain is similar to hydrostatic compression. Moreover (100) lattice planes are equally spaced, as shown in Fig. 3.3a. Let the two atoms in each unit cell be centered at $(0, 0, 0)$ and $(a/4, a/4, a/4)$; then this spacing is just $a/4$. The planes of atoms belong alternately to each sublattice, and the A–B and B–A spacings are equal before and after the strain is applied.

Suppose now that the strain is applied parallel to a (111) bond. Before the strain the (111) lattice planes have an A–B sandwich structure, so that they are not equally spaced; see Fig. 3.3b. The A–B spacing is $(3^{1/2})a/4 = r_{AB}$ but the B–A spacing is $r_{BA} = r_{AB}/3$. Thus the strain need not satisfy the condition for uniform strain

$$\delta r_{AB}/r_{AB} = \delta r_{BA}/r_{BA} \tag{3.26}$$

as it did by symmetry in the (100) case.

One may consider two extreme cases which are illustrated in Fig. 3.4. If there are no bond-bending forces, the internal strain is such that all bond

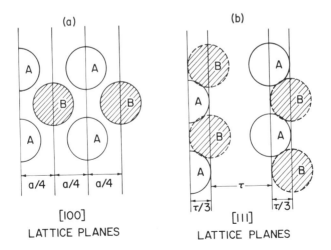

Fig. 3.3. (a) Spacing of [100] lattice planes of A and B ions in zincblende crystals, and (b) spacing of [111] lattice planes.

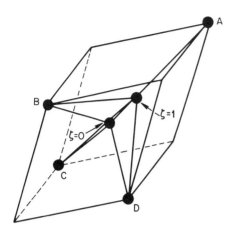

Fig 3 4. Nature of the internal strain parameter ζ for tetrahedrally coordinated atoms. For $\zeta = 0$ the central atom remains at the center of the distorted tetrahedron, which has been stretched along the OA axis. For $\zeta = 1$ the central atom deforms in such a way as to keep the OA bond length equal to the OB, OC, and OD bond lengths.

lengths remain equal after the strain is applied. In the diamond lattice the internal strain parameter ζ defined by (3.18) reduces to

$$\zeta = (\alpha - \beta)/(\alpha + \beta), \qquad (3.27)$$

where α and β are the bond-stretching and bending force constants used in (3.13)–(3.15). Thus $\beta = 0$ corresponds to $\zeta = 1$. On the other hand, one can show that $\alpha = \beta$ and $\zeta = 0$ corresponds to a uniform strain which satisfies Eq. (3.27), while $\zeta < 0$ corresponds to having the strain of the (111) bond greater than the macroscopic strain itself, which is unphysical. Thus one has the inequalities $0 \leq \zeta \leq 1$ and $0 \leq \beta \leq \alpha$.

One can measure ζ directly in diamond-type crystals by X-ray diffraction [4]. If $\zeta > 0$ the strain is nonuniform and certain Bragg reflections which were forbidden in the unstrained crystal are allowed in the presence of strain. In both Si and Ge the X-ray data give $\zeta = 0.63 \pm 0.04$. The values of α and β obtained from the elastic constants yield from (3.27) that $\zeta = 0.54$ in both cases. Thus from the elastic constants one slightly overestimates β in Si and Ge. This is because the α–β model omits longer range forces that are discussed in the next chapter.

PIEZOELECTRIC CONSTANTS

In zincblende crystals the A and B lattice planes in general carry equal and opposite net charges. Thus a nonuniform strain displaces the A sublattices more than the B sublattices and generates an internal electric field. Both the piezoelectric effect and the internal strains just discussed involve tensors of the third rank [5]. Note that a crystal in the shape of a cylinder with major axis along [111] will have a [$\overline{1}\overline{1}\overline{1}$] face composed of A cations and a [111] face composed of B anions. The sign of the piezoelectric field is determined by orienting the applied strain and the internal electric field relative to this axis.

The relationship between the total electric displacement $D_{[111]}$ and the applied strain $S_{[111]}$ is given by [6]

$$D_{[111]} = \tfrac{2}{3}e_{pol}S_{[111]}, \qquad (3.28)$$

where e_{pol} is the electric field produced by [111] bonds alone, while $D_{[111]}$ includes [$1\overline{1}\overline{1}$] bonds, etc. Values [6] of e_{pol} are listed in Table 3.2. For comparison, with $\zeta \approx 0.6$ and a fixed charge $\pm e$ on the A and B sublattices, respectively, one obtains $e_{pol} \approx +0.6$ in the units of Table 3.2.

ORIGIN OF PIEZOELECTRIC EFFECTS

The first thing to note in Table 3.2 is that the piezoelectric coefficients reverse sign on going from I–VII and II–VI to III–V compounds. The signs in the two cases correspond to A cations being net positively charged for A having valence 1 or 2, but net negatively charged for A having valence 3. One can understand this sign reversal qualitatively in terms of the extreme ionic limit (where the A atoms carry a net positive charge equal to their valence Z) or the extreme covalent limit (where the A atoms carry a net negative charge equal to $4 - Z$). For $Z = 2$ one is nearer the ionic or closed shell limit, while for $Z = 3$ one is nearer the covalent or open shell limit.

To understand piezoelectric constants quantitatively one needs a microscopic model to describe the internal electric fields generated when a uni-

TABLE 3.2

The Piezoelectric Coefficient e_{pol} in Zincblende and Wurtzite Crystals.

Crystal[a]	$e_{pol}(C/m^2)$	Crystal[a]	$e_{pol}(C/m^2)$
GaSb	-0.22	BeO	0.04
GaAs	-0.28	ZnS	0.25
InSb	-0.12	CdTe	0.05
InAs	-0.08	ZnSe	0.09
GaP	-0.17	CdS	0.44
AlSb	-0.12	CuCl	0.50
ZnTe	0.05		

[a] The crystals are listed in order of increasing ionicity.

axial strain is applied parallel to a [111] axis of the crystal [7]. Two general effects must be considered:

1. If the ions carry a fixed charge $\pm e_0^*$, the internal strain proportional to ζ will give rise to relative displacement of the $[\overline{111}]$ A atom cation face relative to the [111] B atom anion face. This will generate an internal electric field, and will make a contribution to e_{pol} which is positive.

2. If the ionic charges themselves change with strain, i.e., if

$$e^*(b) = e_0^*(b/b_0)^d, \tag{3.29}$$

where b and b_0 are the strained and unstrained bond lengths parallel to [111], this will also generate an internal electric field, of sign opposite to that of (1). The power d is called the charge redistribution index. Combining (1) and (2) gives for the piezoelectric term [7]

$$e_{pol} = (\zeta - d/4)\,(3^{1/2})\,(e_0^*/a^2). \tag{3.30}$$

The more covalent the crystal, the larger the magnitude of d should be, while conversely the more ionic the crystal, the more valid a rigid ion model corresponding to fixed charges with $d = 0$. We therefore show in Fig. 3.5

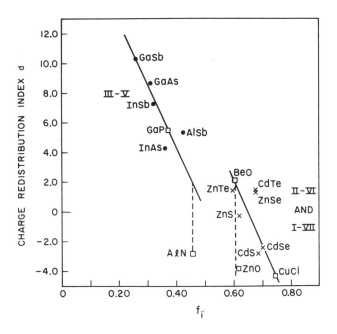

Fig. 3.5. Chemical trends in the charge redistribution indexes which contribute to the piezoelectric coefficient as indicated by Eq. (3.25).

the charge redistribution index d plotted against the spectroscopic ionicity f_i. Linear relations are obtained:

$$d = -43f_i + 21.5, \qquad f_i < 0.6$$

$$d = -43(f_i - 0.15) + 21.5, \qquad f_i > 0.6. \tag{3.31}$$

The small break in linearity between $A^N B^{8-N}$ crystals with $N = 2$ and $f_i > 0.6$ and $N = 3$ with $f_i < 0.6$ might have several explanations. One of these is that the $N = 2$ compounds are significantly closer to the "closed shell" or rigid ion $A^{2+}B^{2-}$ limit than are the $N = 3$ compounds, and this represents an extra effect under pressure. In any case, the dominant effect is the dependence of charge redistribution on bond ionicity. It is perhaps worth noting that the description of the charge transfer associated with bond stretching and bending through (3.29) is oversimplified. The change in bond angles with strain which is illustrated in Fig. 3.4 causes changes in the hybridized directed orbitals which bond each atom to its neighbors, and this redistribution of charge is what is described schematically by (3.29).

WURTZITE CRYSTALS

In this chapter we have discussed second-rank stress and strain tensors, third-rank internal strain and piezoelectric tensors, and fourth-rank elastic tensors. The independent components of these tensors can be determined by group theory [5]. For cubic crystals the independent components can be written down by inspection. But for wurtzite structures there are additional relations which are approximately (but not exactly) valid. These relations are not obtainable by group theory, but they can be derived as follows [8].

The wurtzite structure, as we recall from Chapter 1, has hexagonal symmetry, and consists of two interpenetrating hexagonal close-packed lattices, and is analogous to the zincblende structure, which consists of two interpenetrating face-centered cubic lattices. If the ratio of spacing of lattice planes along the z axis to that normal to it, denoted by c/a, is close to ideal ($c/a = 1.633$), then the nearest and second-nearest-neighbor configurations in the two structures are nearly the same. (See Table 3.3). We may therefore choose to regard both structures as composed of tetrahedral building blocks, neglecting interactions with distant neighbors.

The difference between the two structures lies in the orientation of tetrahedra. In the cubic crystals the tetrahedral corners lie along the [111] directions of the crystal. In the hexagonal structure there are two tetra-

TABLE 3.3

Values of the c/a Axial Ratio for Semiconductors
Having the Wurtzite Structure[a]

Semi-conductor	c/a	Semi-conductor	c/a
AlN	1.600	CdS	1.623
GaN	1.625	CdSe	1.630
InN	1.611	MgS	1.62
BeO	1.623	MgSe	1.622
ZnO	1.602	AgI	1.635
ZnS	1.637	CuH	1.59
ZnSe	1.634	CdTe	1.637
CuBr	1.640	SiC	1.641
CuCl	1.642	BN	1.645
CuI	1.645	ZnTe	1.645
MgTe	1.622		

[a] Each atom is perfectly tetrahedrally coordinated at $c/a = 1.633$. Data from P. Lawaetz, *Phys. Rev.* **5B**, 4039 (1972).

hedral units with one bond along (001), placed so that the orientation of the remaining bonds differs by rotation by π around the z axis. These two may be obtained from a (111) tetrahedron by multiplication of the coordinates of the latter by appropriate rotation matrices R_{ij} and S_{ij}, respectively.

For a tensor of given rank, one now establishes the independent components of the cubic tensor and then uses these to generate the (approximate) components of the wurtzite tensor. As an example, consider third-rank tensors. The only independent component of a third-rank tensor α_{ijk} consistent with tetrahedral symmetry is α_{123}. In the hexagonal crystal α_{ijk} can be obtained from α_{pqr} of the cubic crystal by

$$\alpha_{ijk} = \tfrac{1}{2} \sum_{pqr} (R_{ip}R_{jq}R_{kr} + S_{ip}S_{jq}S_{kr})\alpha_{pqr}. \tag{3.32}$$

From (3.28) one can express all three independent components α_{333}, $\alpha_{311} = \alpha_{322}$, and $\alpha_{113} = \alpha_{223}$ of α in the wurtzite structure in terms of α_{123} in the zincblende structure. This yields the relations

$$\alpha_{311} = \alpha_{113}, \tag{3.33}$$

$$\alpha_{311} = \alpha_{333}/2. \tag{3.34}$$

This method can also be used for fourth-rank tensors to reduce the number of independent elastic constants. Were it not for the effects of internal strain, in wurtzite one would have $c_{33} = c_{11}$; in actual crystals, e.g., CdS, $c_{33}/c_{11} = 1.09$. With Robinson's model, however, Martin has derived "effective" cubic elastic constants c_{11}, c_{12}, and c_{44} for wurtzite crystals. The reduced shear modulus $(c_{11} - c_{12})/2c_0$ in the Keating-Martin model is nearly proportional to β/α. This is shown in Fig. 3.6 for both sphalerite and wurtzite crystals. The trend for the Cu salts as the bond ionicity f_i tends to the critical ionicity 0.785 may be indicative of the approaching phase transition to the NaCl structure.

Another way of describing the difference between wurtzite and sphalerite structures is to look at the conformation of two tetrahedral units A_3B and

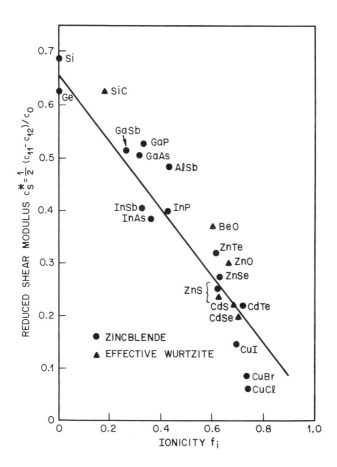

Fig. 3.6. Chemical trends in the reduced shear modulus $c_s{}^*$ as a function of spectroscopic ionicity f_i for zincblende and wurtzite compounds.

AB$_3$ in the two structures, with the BA axis along a (111) body diagonal in the sphalerite structure, and along the (001) axis in the wurtzite structure. The eight atoms of the two tetrahedra are structurally analogous to the ethane molecule H$_3$C–CH$_3$. If one looks along the C–C axis of ethane, one sees that there are two possible conformations for the two H$_3$ groups, vertically above each other (the eclipsed conformation), and rotated by π relative to one another (the staggered conformation). Predominantly covalent bonding always favors the staggered conformation, which is why crystals with low ionicity favor the cubic sphalerite structure. However, as the ionicity increases, the Coulomb attraction between A$_3$ and B$_3$ favors the eclipsed conformation, which is why many II–VI compounds, such as CdS, have the wurtzite structure.

CHALCOPYRITE CRYSTALS

If a unit cell of the sphalerite structure is doubled along the z axis, it will contain four atoms, two cations and two anions. The anions play the greater role in stabilizing the structure because they contain most of the valence electrons. Therefore, it is possible to replace the two cations X^N in a X$_2^N$C$_2^{8-N}$ sphalerite crystal with cations A^{N-1} and B^{N+1} to obtain a crystal of formula A^{N-1}B^{N+1}C$_2^{8-N}$. In some compounds the A and B cations are randomly distributed on the face-centered cubic cation sublattice. However, the A and B cations may also form an ordered array of centered 2 × 2 units in xy planes, with each A atom at the center of a square of B atoms, and vice versa. In the xz and yz planes the arrays will have less symmetrical forms with two adjacent corners of the square occupied by A atoms, the other two corners by B atoms. Because of this the z axis is distinguished from the x and y axes, and the c/a ratio of the resulting tetragonal structure will in general differ from the ideal value of 2.000 that holds when the distribution of A and B cations on the cation sublattice is a random one.

The experimental values of c/a for a number of crystals with this structure, which is called chalcopyrite, are shown in Table 3.4. In all cases it is found that $2 - c/a > 0$. Moreover, the departures of c/a from the ideal value are much greater for the chalcopyrite structure than for the wurtzite structure (of order 5% compared to 1%). The reason for this is easy to understand: in wurtzite the distortion is caused by third-neighbor interactions, whereas in chalcopyrite the distortion is caused by second-neighbor interactions.

In wurtzite crystals c/a is found to lie both above and below the ideal value by amounts of order 1%, which shows that c/a is determined by details of the atomic potentials, details which are not dependent simply on atomic size or valence. In A^{N-1}B^{N+1}C$_2^{8-N}$ chalcopyrite crystals, on the other

TABLE 3.4

Tetragonal Distortions in Chalcopyrite Crystals at
Room Temperature

Compound	$2 - c/a$	Compound	$2 - c/a$
$CuAlS_2$	0.04	$AgAlS_2$	0.20
$CuGaS_2$	0.04	$AgGaS_2$	0.21
$CuInS_2$	0.00	$AgInS_2$	0.08
$CuAlSe_2$	0.06	$AgAlSe_2$	0.20
$CuGaSe_2$	0.04	$AgGaSe_2$	0.18
$CuInSe_2$	0.00	$AgInSe_2$	0.08
$CuTlSe_2$	0.01		
$CuAlTe_2$	0.03	$AgAlTe_2$	0.12
$CuGaTe_2$	0.01	$AgGaTe_2$	0.10
$CuInTe_2$	0.00	$AgInTe_2$	0.04
$ZnSiP_2$	0.067	$ZnSiAs_2$	0.057
$ZnGeP_2$	0.040	$ZnGeAs_2$	0.034
$ZnSnP_2$	(disordered)	$ZnSnAs_2$[a]	0.000
$CdAiP_2$	0.163	$CdSiAs_2$	0.152
$CdGeP_2$	0.123	$CdGeAs_2$	0.112
$CdSnP_2$	0.05	$CdSnAs_2$	0.043
$MgSiP_2$	0.232		

[a] The compound $ZnSnAs_2$, which shows no tetragonal distortion, can also be prepared in the sphalerite structure where the Zn and Sn are distributed in a disordered fashion on the cation sublattice.

hand, $2 - c/a$ is always positive and may be as large as 0.1–0.2. The positive sign may be understood by noticing that each A atom experiences second-neighbor interactions only with B atoms in the xy plane, but with equal numbers of A and B atoms in the xz and yz planes. Therefore, $2 - c/a > 0$ means that the strengths of A–A and B–B interactions are increased at the expense of A–B interactions. This probably reflects the "softness" of either the A–C–A or B–C–B bond-bending linkage compared to the A–B–C linkage. This softness reflects also the destabilizing effects of a deficiency or excess of bonding electrons in the A–C–A or B–C–B linkages, respectively, compared to the A–B–C linkage, which averages eight electrons per cation–anion pair.

SUMMARY

Interatomic forces in semiconductors are best described by valence force fields, which include bond-bending and bond-stretching forces, as well as combinations thereof. Only nearest-neighbor bond-stretching forces are of the central type found in ionic crystals. Bond-bending forces are directional, and are characteristic of covalent bonding. The ratio of directional to central, or bending to stretching, forces characteristically decreases linearly with increasing spectroscopic ionicity f_i of the bond.

REFERENCES

1. C. Kittel, "Introduction to Solid State Physics," Wiley, New York, 1953, 1956, 1966.
2. P. N. Keating, *Phys. Rev.* **145**, 637 (1966).
3. R. M. Martin, *Phys. Rev.* B **1**, 4005 (1970).
4. A. Segmuller and H. R. Neyer, *Phys. Kondens. Mater.* **4**, 63 (1965).
5. J. F. Nye, "The Physical Properties of Crystals." Oxford Univ. Press (Clarendon), London and New York, 1960.
6. G. Arlt and P. Quadflieg, *Phys. Status Solidi* **25**, 323 (1968).
7. J. C. Phillips and J. A. VanVechten, *Phys. Rev. Lett.* **23**, 1115 (1969).
8. F. N. H. Robinson, *Phys. Lett.* **26A**, 435 (1968).

4

Lattice Vibrations

The sound waves discussed in the preceding chapter represent special cases of harmonic waves in crystals, namely the limit in which the wavelength $\lambda \to \infty$ or the wave number $k \to 0$. In this limit the frequency is linearly related to the velocity of sound, $\omega = ck$. When λ is reduced to the point that it becomes comparable to the lattice constant a, dispersion sets in and usually $\omega < ck$. For example, one might have for k along a symmetry axis,

$$\omega = \omega_0 \sin(ck/\omega_0). \qquad (4.1)$$

It follows that knowledge of $\omega(k)$, for $k \sim \pi/a$, can give us more information on interatomic forces. In particular the elastic constants and the velocity of sound involve certain weighted averages of short-range and long-range forces. As λ decreases towards a, the effect of the long-range forces tends to cancel out because of the oscillatory character of wave motion. Thus $\omega(k)$ at large k (small λ) gives more information about short-range forces.

BRILLOUIN ZONES

Because of the periodicity of the crystal lattice, apart from a phase factor the amplitudes of one type of lattice vibration or one type of one-electron wave function are the same in each unit cell. The difference between the phase factor for the unit cell centered at R_m and the one centered at R_n is $\exp[i\mathbf{k}\cdot(R_m - R_n)]$. Just as one can construct unit cells in real space, so there are unit cells in $\mathbf{k}$ space. One may also introduce $\mathbf{p} = \hbar\mathbf{k}$ as the crystal momentum. If $\mathbf{R}$ is a primitive lattice vector in real space, then $\mathbf{G}$ is a primitive lattice vector in $\mathbf{k}$ space, sometimes called reciprocal space. For each $\mathbf{R}$ one defines $\mathbf{G}$ by

$$\mathbf{G}\cdot\mathbf{R} = 2\pi. \qquad (4.2)$$

The smallest unit cell in reciprocal space is called the first Brillouin zone. All $\mathbf{k}$ lying inside the first Brillouin zone are inequivalent in the sense that if $\mathbf{k}_1$ lies inside the first zone, then subtracting any $\mathbf{G}$ from $\mathbf{k}_1$ does not give a $\mathbf{k}_2$ which is also inside the zone. The second Brillouin zone consists of points $\mathbf{k}_2$ such that $\mathbf{k}_2 = \mathbf{k}_1 + \mathbf{G}$, while the third zone contains points $\mathbf{k}_3$ such that $\mathbf{k}_3 = \mathbf{k}_1 + \mathbf{G} + \mathbf{G}'$, and so on.

Brillouin zone faces have special symmetry properties associated with the periodicity and symmetry of the reciprocal lattice. For example, along the symmetry axis where an acoustic wave is described by (4.1), the group velocity $v_g(k)$ is given by

$$v_g(k) = d\omega/dk = c\cos(ck/\omega_0).$$

As $k \to 0$, $v_g(k) \to c$, the sound velocity. As $k \to \pi\omega_0/2c$, $v_g(k) \to 0$; i.e., we have standing waves. This frequently happens at Brillouin zone faces because the faces have mirror symmetry normal to the face.

When there is only one type of normal mode, one plots the frequencies of that mode as a function of $\mathbf{k}$ in the first Brillouin zone. In a crystal with s atoms per unit cell there are $3s$ normal modes. The lattice wave associated with each kind of normal mode is completely determined once $\mathbf{k}$ is known modulo $\{\mathbf{G}\}$, where $\{\mathbf{G}\}$ is the set of reciprocal lattice vectors. One could therefore plot the $3s$ normal mode frequencies in the first $3s$ Brillouin zones. However, the geometry of Brillouin zones beyond the first one is generally quite complicated. It is therefore customary to plot all the normal modes in the first Brillouin zone, labelling them according to the kind of normal mode involved.

We have seen in our discussion of crystal structures of semiconductors in Chapter 1 that most common semiconductors crystallize in either diamond, zincblende, or (as for Pb salts) NaCl structures. All three structures correspond to face-centered cubic arrays of unit cells, although the arrange-

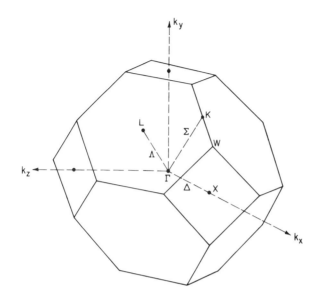

Fig. 4.1. The Brillouin zone of face-centered cubic lattices. Principal symmetry points and lines are labeled.

ment of atoms within each unit cell is different. Because the lattice which is reciprocal to face-centered cubic is body-centered cubic, the first Brillouin zone for these structures is just the unit cell of the body-centered cubic lattice. This is shown in Fig. 4.1 together with the letters conventionally used to label each point or line of definite symmetry in **k** space. In Table 4.1 these points are given in Cartesian coordinates.

TABLE 4.1

Cartesian Coordinates of Symmetry Points and Lines in the Face-Centered Cubic Brillouin Zone Shown in Fig. 4.1

Symbol	Coordinates[a]	Symbol	Coordinates[a]
Γ	$(0,0,0)$	W	$(2,1,0)$
X	$(2,0,0)$	Δ	$(2x,0,0)$
L	$(1,1,1)$	Λ	(x,x,x)
K = U	$(\frac{3}{2},\frac{3}{2},0)$	Σ	$(\frac{3}{2}x,\frac{3}{2}x,0)$

[a] Here x denotes a number between 0 and 1, and the coordinates are given in units of π/a where a is the cubic lattice constant.

In discussing lattice vibrations and energy bands we shall frequently have occasion to refer to definite symmetry points or lines. Of course every point in **k** space has the same statistical weight as every other point, so that when we refer to a single point it will be with the understanding that the physical property under discussion depends on the properties of lattice vibrations or electronic states associated with a neighborhood of the given point. The extent of this neighborhood is usually determined by the energy range of interest, which for electronic states may be $k_B T$, for instance. A typical electronic bandwidth in a semiconductor is 1 eV, so that the nature of the normal modes of interest may vary slowly over the range $k_B T \approx$ 0.025 eV at room temperature, and will almost always vary negligibly at hydrogen or helium temperatures. Corrections to point symmetry properties are, however, known in exceptional cases where the energy gap to another normal mode is not large compared to $k_B T$.

EXPERIMENTAL DETERMINATION OF ω(k)

The determination of $\omega(\mathbf{k})$ in the short-wavelength regime long posed a problem. The typical frequency, $\omega \sim 10^{13}$ sec^{-1}, is too high to be generated by ultrasonic transducers. Some information about the overall frequency spectrum $dN(\omega)/d\omega$ can be obtained by studying specific heats. The problem of inverting a given specific heat curve $c_v(T)$ to obtain $dN(\omega)/d\omega$ has never been solved analytically, even if one supposes that anharmonic contributions to the experimental values of $c_v(T)$ can be neglected. However, one can choose a particular model and adjust the parameters to fit experiment. In this manner there has existed for several decades a thriving cottage industry based on model coupled oscillator lattices. From the models with very precise values of $c_v(T)$ one could sometimes determine $\omega(\mathbf{k})$ quite well, at least for the lower-lying branches of the spectrum.

In the late 1950s much more complete knowledge of lattice vibration spectra became available through neutron studies. In principle a beam containing neutrons of definite energy $\hbar\omega$ and momentum $\hbar\mathbf{k}$ is incident on a large single crystal. In addition to elastic Bragg scattering from lattice planes of the crystal, the neutrons may also be scattered inelastically to a new energy $\hbar\omega'$ and new momentum $\hbar\mathbf{k}'$. A phonon of energy $\hbar(\omega - \omega')$ and momentum $\hbar(\mathbf{k}' - \mathbf{k})$ is absorbed or emitted, and after many measurements one determines $\omega(\mathbf{q})$, where $\mathbf{q} = \mathbf{k}' - \mathbf{k}$, for values of $\mathbf{q}$ along symmetry axes of the crystal. These then generate a quite complete picture of lattice dispersion curves. At present the accuracy of the experimentally determined dispersion curves $\omega(\mathbf{q})$ for Si and Ge may be as good as 3%, which is greater than the significance of the microscopic models which theorists have constructed to investigate interatomic forces.

NORMAL MODES

Suppose there are M atoms per unit cell, labeled by $m = 1, \ldots, M$. In a lattice wave the mth atom in the unit cell labeled by $\mathbf{R}_l$ is displaced by $u_a{}^m \exp(i\mathbf{q}\cdot\mathbf{R}_l - i\omega t)$. Substitution of this expression in Newton's law gives simultaneous equations for the amplitudes $u_a{}^m$ in terms of second derivatives of the lattice potential energy. (These are denoted in Eq. (3.6) by the coefficients $K_{aa'}^{mm'}$). When interactions between atoms in different unit cells are included, the same amplitudes $u_a{}^m$ reappear, weighted by a phase factor. Thus there are only $3M$ independent amplitudes, giving rise to $3M$ independent normal modes. The effective force constants for a lattice wave of wave vector $\mathbf{q}$ are

$$K_{aa'}^{mm'} \text{ (effective)} = \sum_l K_{aa'}^{mm'} \exp(i\mathbf{q}\cdot\mathbf{R}_l). \qquad (4.3)$$

Thus the problem of interactions among all the atoms in the crystal is reduced to a problem of M atoms interacting through a harmonic potential with the effective force constants (4.3).

Among the $3M$ normal modes, three of them, two transverse and one longitudinal, have the property that as $q \to 0$, $\omega = cq$. These are the elastic waves studied in the preceding chapter, for which we have information about c accurate to about 0.1%. For larger q these are called acoustic modes. The remaining $3M - 3$ modes are called optic modes. At $q = 0$ the optic modes correspond to internal vibrations of the sublattices against each other which leave the center of mass of the unit cell unchanged. Thus if the mass of ion m is μ_m, the atomic displacements $u_a{}^m$ in a $q = 0$ optic mode satisfy the relation

$$\sum_m \mu_m u_a{}^m = 0. \qquad (4.4)$$

In order to determine what information about the crystal structure these modes contain, one must usually turn to microscopic models. There are, however, a few general remarks which can be made.

MODE DESCRIPTIONS

Among the three acoustic modes, the longitudinal mode (displacement vectors parallel to wave vector) has the highest frequency, except possibly at very short wavelengths (k near a Brillouin zone face). Longitudinal modes emphasize bond-stretching, whereas transverse modes (displacement vectors perpendicular to wave vector) emphasize bond-bending, as described by the force constants α and β in the preceding chapter. (The arguments comparing macroscopic and microscopic strains showed $\beta < \alpha$, which also makes the transverse modes have lower frequency). These modes are frequently abbreviated by "la" and "ta," for longitudinal acoustic and transverse acoustic, respectively.

In a diatomic crystal there are six normal modes, three acoustic, and three optic. The optic modes are abbreviated by "lo" and "to," for longitudinal optic and transverse optic, respectively. The optic modes often all have about the same frequency, independent of polarization and wave vector, although the longitudinal modes usually have somewhat higher frequencies. In partially ionic crystals, Coulomb forces increase "lo" frequencies, relative to "to" frequencies, as we shall see below.

The stability of a given crystal is indicated by the ratio of transverse to longitudinal frequencies. When the overall structure changes, this shows up as a reduction in "ta" compared to "la" frequencies. This is similar to the softening of the shear elastic constants compared to the hydrostatic ones, as discussed in connection with Fig. 3.2. Another possible structural change involves internal displacements of atoms in the unit cell relative to one another without changing unit cells. Such internal displacements usually involve internal shears and are describable in terms of "to" modes. Indeed, the "to" mode frequency may tend to zero, although it will not quite reach zero because of anharmonic forces. If the internal displacement involves charged ions, the unit cell may develop a permanent electric dipole moment; i.e., the crystal becomes ferroelectric. Behavior of this kind has been observed in $SrTiO_3$ and $BaTiO_3$.

SUM RULES

One may divide the effective force constants $K_{aa'}^{mm'}$ into two groups, corresponding to nearest-neighbor forces and forces associated with non-nearest-neighbors. One may call the latter medium and long-range forces. The force of longest range is of course the Coulomb interaction. In most covalent crystals nearest-neighbor atoms belong to different sublattices; e.g., in $A^N B^{8-N}$ semiconductors the nearest-neighbors of the A atoms are B atoms, and vice versa.

Nearest-neighbor forces are associated either with overlap of hybridized valence orbitals or with closed-shell repulsion, for which no general statement can be made. These forces are, however, off-diagonal ($m' \neq m$). The diagonal contributions to the force constant matrix are represented by

K_{aa}^{mm}. Suppose we take the trace of this matrix; it must be

$$\sum_{a,m} K_{aa}^{mm} = \sum_i \omega_i^2(\mathbf{q}). \tag{4.5}$$

Now suppose that the medium and longer range forces K_{aa}^{mm} are all electrostatic in origin, i.e., can be described in terms of interactions between nonoverlapping electrostatic multipoles. Each of the matrix elements K_{aa}^{mm} depends on $\mathbf{q}$, but for electrostatic interactions their sum does not. To see this, consider the contributions to K_{aa}^{mm} (eff.) from one particular atom of type m. The contribution is proportional to the second derivative with respect to x_a of the potential between the special atom we are considering and another one on the same sublattice, and the sum of this over the three a's vanishes because electrostatic potentials obey Laplace's equation. Thus (4.5) is independent of $\mathbf{q}$ and we have the Brout–Rosenstock sum rule [1]

$$\sum_i \omega_i^2(\mathbf{q}) = \text{const.} \tag{4.6}$$

The validity of the sum rule (4.6) for $\mathbf{q}$ parallel to the $\langle 111 \rangle$ bond directions is tested in Fig. 4.2. For Si and Ge the rule is obeyed quite well, but for diamond it fails badly.

If one considers this result in terms of the valence force fields described in the previous chapter, then it is clear that the bond-bending forces

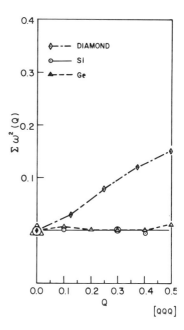

Fig. 4.2. The sum rules for lattice vibration frequencies, Eq. (4.6), is tested for $\mathbf{q}$ along the Λ or [111] axis.

involve interactions which depend on the angle $\varphi = \langle A_1BA_2 \rangle$; i.e., bond-bending forces represent a nonelectrostatic interaction between the second neighbors A_1 and A_2. Thus bond-bending forces violate (4.6). From the elastic constants we concluded that these forces, represented by the parameter β, were quite strong in diamond compared to the bond-stretching forces, represented by α. In Si and Ge β/α was found to be much smaller than in diamond. Of course, the elastic constants represent three numbers known to 0.1%, while the dispersion curves (which are also measured in $\langle 100 \rangle$ and $\langle 110 \rangle$ directions) provide nearly 20 values of (4.3) accurate to 3%. Thus the entire dispersion curves probably contain more information than the elastic constants. From them we conclude that valence force fields are useful for describing the short wavelength lattice vibrations of diamond but not Si or Ge. For the latter crystals the medium and long-range forces are more nearly electrostatic in nature.

OPTICALLY ACTIVE MODES

There are two kinds of lattice vibration modes which interact with light. In both cases the wavelength λ of light in the infrared and visible portions of the spectrum is thousands of angstroms or more, and therefore λ is much greater than the dimensions of a unit cell in the crystal which is of order a few angstroms. Thus the crystal momentum or wave number $k = 2\pi/\lambda$ is almost zero compared to a reciprocal lattice vector G. For most purposes one can take $k = 0$, although in the presence of Coulomb forces one must carefully pass to the limit $k \to 0$ after the dimensions of the crystal have been allowed to go to ∞.

INFRARED MODES AND EFFECTIVE CHARGES

If a lattice vibration is polarized transverse to its direction of propagation and has the transformation properties of a vector (after summation over all groups of equivalent atoms in the unit cell), that vibration may interact with an electromagnetic wave described by $(\mathbf{E}, \mathbf{H})$ through an electric dipole coupling d, of the form

$$d_1 = \sum_i e_i^* \mathbf{u}_i \cdot \mathbf{E}_i. \tag{4.7}$$

As examples of infrared-active modes, consider the optic modes in Ge and in GaAs. In Ge the effective charges e_i^* for the two Ge atoms in the unit cell must be equal, but as the crystal is neutral, their sum must be zero. Thus $e_i^* = 0$ and the optical modes in Ge are infrared-inactive.

In GaAs, on the other, hand charge neutrality requires $e^*(\text{Ga}) = e^* = -e^*(\text{As})$ and because the nuclear masses are approximately equal, $\mathbf{u}(\text{Ga}) = \mathbf{u} = -\mathbf{u}(\text{As})$. Thus (4.7) becomes $d_1(\text{GaAs}) = 2e^*\mathbf{u}\cdot\mathbf{E}$ and the transverse optic modes are infrared-active.

The interpretation of e^* in (4.7) is straightforward if we identify $\mathbf{E}_i$, the internal field which couples to the transverse optic modes of vibration, with $\mathbf{E}$, the macroscopic field of the external electromagnetic wave. This is the usual procedure in semiconductors. It neglects the classical Lorentz local field corrections which are important in gases or liquids, and which arise because of interaction between localized polarizable centers.

The effective charge e^* which occurs in (4.7) measures the interaction of the dynamical lattice transverse optic mode with the transverse electromagnetic field. It is sometimes denoted by e_T^*. Another definition of e^* which allows for the polarization of the lattice by longitudinal electric fields is given in Eq. (4.12). The two are related by $(e_T^*)^2 = \epsilon_0(e^*)^2$. Values of e^* for tetrahedrally coordinated semiconductors are listed in Table 4.2.

RAMAN ACTIVE MODES

The process of infrared absorption creating a transverse optical phonon is shown diagramatically in Fig. 4.3a. Also shown in Fig. 4.3b is the more

TABLE 4.2

Values of e^* for Tetrahedrally Coordinated
Semiconductors[a]

Crystal	$|e^*|$	Crystal	$|e^*|$
BN	0.55	GaN	0.42
SiC	0.41	AlN	0.40
BP	0.10	BeO	0.62
AlP	0.28	ZnO	0.53
AlSb	0.19	ZnS	0.41
GaP	0.24	ZnSe	0.34
GaAs	0.20	ZnTe	0.27
GaSb	0.15	CdS	0.40
InP	0.27	CdSe	0.41
InAs	0.22	CdTe	0.34
InSb	0.21	CuCl	0.27

[a] These values include no local field corrections and they satisfy Eq. (4.12).

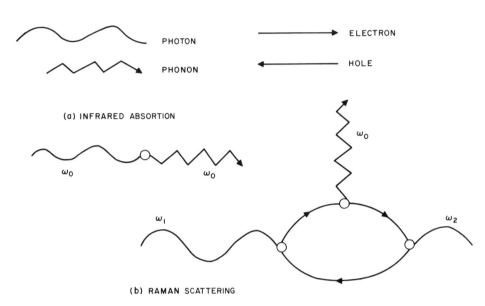

Fig. 4.3. Diagrams illustrating (a) infrared absorption and (b) Raman scattering.

complicated process called Raman scattering. In this process an incoming photon of frequency ω_1 is absorbed inelastically, creating a virtual electron–hole pair. Either the electron or hole interacts with the lattice, creating or destroying an optical phonon. The electron and hole recombine to emit a scattered photon of frequency ω_2, whose energy differs from that of the original photon by the energy of the transverse optic phonon.

The polarizability of the crystal as seen by the initial photon at ω_1 differs from that seen by the final photon at ω_2. This change in polarizability is partly produced by the phonon which has been created or destroyed. One is accustomed to thinking of the electronic states of the crystal in terms of lattice of atomic cores at rest. When the latter are vibrating, however, the electronic states must be calculated in the spirit of the Born–Oppenheimer or adiabatic approximation, in which the electronic motion follows that of the ions. Different ionic configurations will then generate different polarizabilities, making Raman scattering possible. The interaction energy d_2 in this case has the form

$$d_2 = \sum_{\alpha\beta} P_{\alpha\beta}(\mathbf{k}_1, \omega_1; \mathbf{k}_2, \omega_2) E_{1\alpha} E_{2\beta} \tag{4.8}$$

where α and β are Cartesian indices. The polarizability in (4.8) is a function of the polarizations and frequencies of the incident and scattered photons.

POLARITONS

For infrared absorption, conservation of energy requires that the frequency of the light equals that of the transverse optic mode. For Raman scattering (4.8) shows that ω_1 may have any value. Experimentally it has been found that $P(\mathbf{k}_1, \omega_1; \mathbf{k}_2, \omega_2)$ is large in regions where there are excitons or exciton resonances. The details of the coupling, e.g., the differences between longitudinal optic and transverse optic coupling constants, are not well understood at this writing. However, the nature of the strong coupling is known to arise from the mixing of the photon field with the exciton field, as described for photons and phonons by Born and Huang [2]. The mixed exciton-phonon or phonon-photon field is sometimes called a *polariton*, and the way the mixing takes place is shown in Fig. 4.4. The sketch includes for simplicity only the transverse polarization modes of the lattice, the excitons and light waves.

Both the lattice phonons and the excitons contribute to the retardation of electromagnetic waves in the crystal by reducing the group velocity

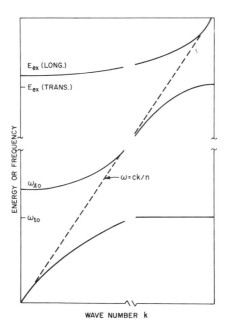

WAVE NUMBER k

Fig. 4.4. A sketch showing how optic phonons mix with photons (lower part of figure) and how excitons mix with photons (upper part). Note the break in scale between the upper and lower parts. Also shown is the dispersion relation for light in the absence of mixing, where n is the index of refraction.

TABLE 4.3

Values of ω_{to} and ω_{lo} in Semiconductors[a]

Crystal	ω_{lo} $\quad$ ω_{to} (10^{12} Hz)		Crystal	ω_{lo} $\quad$ ω_{to} (10^{12} Hz)	
C	39.96	39.96	GaN	24.	16.5
Si	15.69	15.69	AlN	27.5	20.0
Ge	9.02	9.02	InAs	7.30	6.57
Sn	6.0	6.0	InSb	6.00	5.37
SiC	29.15	23.90	BeO	32.7	21.1
BN	40.20	31.95	ZnO	17.4	11.8
BP	25.0	24.6	ZnS	10.47	8.22
AlP	15.03	13.20	ZnSe	7.53	6.12
AlSb	10.20	9.60	ZnTe	6.18	5.47
GaP	12.08	10.96	CdS	9.1	6.9
GaAs	8.76	8.06	CdSe	6.52	5.13
GaSb	7.30	6.90	CdTe	5.13	4.20
InP	10.35	9.10			

[a] Data collected by R. M. Martin (unpublished, 1969).

defined by (4.2). The effect of the transverse optic phonons of frequency ω_{to} is described by the dispersion relation

$$k^2 c^2/\omega^2 = \epsilon_0 + (\epsilon_s - \epsilon_0)/(\omega_{to}^2 - \omega^2) \qquad (4.9)$$

where ϵ_0 is the low-frequency limit of the electronic contribution to the dielectric constant. The lattice contribution to the static dielectric constant ϵ_s is represented by $\epsilon_s - \epsilon_0$. Because of the lattice contribution, there are no solutions to (4.9) in the region

$$\omega_{to} < \omega < \omega_{lo} \qquad (4.10)$$

where ω_{lo} is the longitudinal optic mode frequency given by the relation

$$\omega_l^2 = (\epsilon_s/\epsilon_0)\omega_0^2. \qquad (4.11)$$

In the frequency region (4.10) the crystal would be totally reflecting in the absence of damping. Values of ω_{to} and ω_{lo} are listed in Table 4.3. These are related to the effective charges e^* by [3]

$$\omega_{lo}^2 - \omega_{to}^2 = (4\pi\epsilon_0/M\Omega)(e^*)^2 \qquad (4.12)$$

where M is the reduced mass and Ω the volume of the unit cell. Relations similar to (4.9)–(4.12) can also be derived for exciton–photon interactions.

DISPERSION CURVES OF DIAMOND-TYPE SEMICONDUCTORS

The lattice vibration frequencies $\omega(\mathbf{k})$ of Ge for $\mathbf{k}$ along certain symmetry directions are shown in Fig. 4.5. There is a clear-cut separation of acoustic and optic modes. Compared to other insulators (e.g., NaCl) the only salient feature of the dispersion curves is the flattening of the transverse acoustic mode frequencies as $\mathbf{k}$ approaches the Brillouin zone boundary in the [100] and [111] directions. The flattening is negligible in diamond, but is substantial in Si, and becomes even more marked in Ge and gray Sn.

To illustrate this trend the [100] frequencies of diamond and Si are compared on a reduced scale in Fig. 4.6. The transverse acoustic frequencies in diamond represent a natural extrapolation of the low-frequency behavior, while in Si the flattening is quite evident.

A natural microscopic explanation of this short-wavelength behavior cannot be found in the valence force field model used in the previous chapter to discuss elastic constants. The reason for this is that interactions of an atom at $\mathbf{R} = 0$ with an atom at (X, Y, Z) are modulated by $\cos(k_x X + k_y Y + k_z Z)$. To reproduce the flattening effect at short wave-

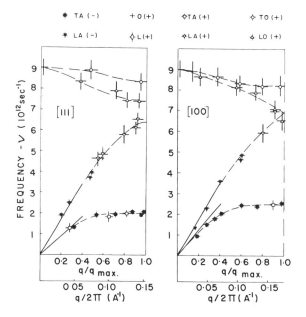

Fig. 4.5. Dispersion curves $\omega(\mathbf{k})$ of Ge as measured by neutron scattering [from B. N. Brockhouse and P. K. Iyengar, *Phys. Rev.* **111** 747 (1958). Courtesy Atomic Energy of Canada, Ltd.].

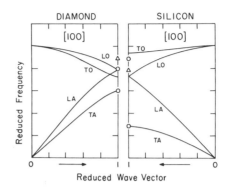

Fig. 4.6. Comparison of diamond and Si[100] vibration frequencies.

lengths one must introduce interactions with many distant neighbors, thereby in effect providing a Fourier synthesis of $\omega(\mathbf{k}) = \text{const.}$ This approach is artificial and does not provide any explanation for the flattening effect itself.

ELECTROSTATIC MODELS

A microscopic explanation of the short-wavelength flattening of $\omega(\mathbf{k})$ based on the known electronic properties of diamond-type crystals has been given [4] and applied to Si in particular by Martin [5]. In some ways the interatomic forces in Si, Ge, gray Sn and other semiconductors not containing first-row atoms resemble those of metals. This is because the average energy gap E_g is so small compared to the valence plasma energy $\hbar\omega_p$ (see Tables 2.1 and 2.2 for examples). According to Eq. (2.13), this means that $\epsilon_0 \gg 1$.

The electronic dielectric constant ϵ_0 describes the screening of ion–ion forces at distances large compared to a nearest-neighbor distance. In metals ϵ_0 is infinite, and as a result the bare ion–ion interaction $(Z_1 e)(Z_2 e)/r$ is multiplied approximately by the Thomas–Fermi screening factor $\exp(-k_s r)$, where k_s is the Thomas–Fermi wave number discussed in the following chapter. In semiconductors the screening factor is just $1/\epsilon_0$, which is also small in Si, Ge, etc.

The factor $1/\epsilon_0$ accounts for most but not all of the valence screening. Because the crystal is neutral there must be a second part of the valence charge which also contributes to interatomic forces. The first part is centered on the atoms, and it consists of screening clouds which are approximately spherically symmetric about each atom. In order to describe the directed valence bonds the second part of the valence charge distribution is represented by point charges of magnitude proportional to $1/\epsilon_0$ centered at bond sites located halfway between nearest-neighbors. The nature of the electronic charge distribution is sketched in Fig. 4.7.

The representation of the bond charges as point charges is a great over-simplification, but it is probably no worse in this respect than the separation of the valence charge distribution into two distinct parts, one atomic and one bond. It has the advantage of defining a simple model which explains a number of the basic trends in interatomic forces and dielectric properties. One of these is the trend of flattening of the transverse acoustic dispersion curves near the Brillouin zone boundary. To normalize the lattice vibration frequencies one plots them in reduced units $\omega(\mathbf{k})/\omega_{to}(0)$, where $\omega_{to}(0)$ is the Raman ($\mathbf{k} = 0$) frequency. In Fig. 4.8 are shown the values of $\omega_{ta}^2(\mathbf{X})/\omega_{to}^2(0)$ for diamond-type crystals. Here $\omega_{ta}(\mathbf{X})$ is the transverse acoustic frequency at $\mathbf{k} = \mathbf{X}$, which is the center of the (100) Brillouin zone face. At this point both the wave vector $\mathbf{k}$ and the polarization vector of the transverse acoustic wave make equal angles with the [111] bonds of the crystal.

One can make detailed calculations [5] of the interatomic forces and show that only for nearest-neighbors do the forces of atom–atom type differ appreciably from the asymptotic limiting value $(Z_1e)(Z_2e)/\epsilon_0 r$. Thus second-neighbor interactions of the bond-bending type, which as we

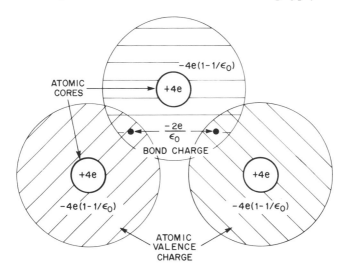

Fig. 4.7. Separation of valence charge distribution in diamond-type semiconductors into atomic and bonding parts.

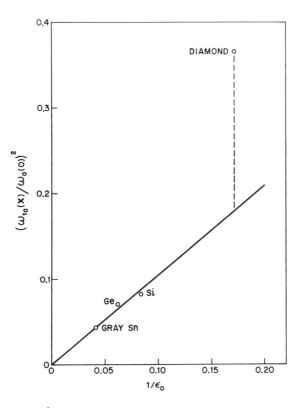

Fig. 4.8. Plot of ω_{ta}^2 $(X)/\omega_{to}^2(0)$ against $1/\epsilon_0$ for diamond-type crystals.

saw in Chapter 2 stabilize tetrahedrally coordinated structures, are associated in the electrostatic model with Coulomb interactions of the atom–bond and bond–bond type. All of these scale with $1/\epsilon_0$. The nearest-neighbor atom–atom force corresponds to α in the valence force field model, and the atom–bond and bond–bond forces, loosely speaking, correspond to β and more distant forces. As the latter stabilize shear modes, it is to be expected that $\omega_{ta}^2(\mathbf{X})/\omega_{to}^2(0)$ should be nearly a linear function of $1/\epsilon_0$. This is shown to be the case for Si, Ge and gray Sn by Fig. 4.8. The deviation in the case of diamond arises because for crystals containing first-row atoms a valence force-field model is more appropriate than an electrostatic model.

The success of the bond–charge model in explaining interatomic forces at very short wavelengths has led to other applications in the field of nonlinear dielectric properties which are discussed in Chapter 7. Calculations of valence charge densities have been made (see Chapter 6) which exhibit well-localized bond charges of the type used here to discuss interatomic forces. The arbitrary character of the separation of the valence charge distribution into atomic and bond parts means that the model should be handled with care. Nevertheless, it does describe an important part of the electronic structure of semiconductors, and it may well find a number of successful applications of a semiempirical nature.

ZINCBLENDE-TYPE DISPERSION CURVES

When the two atoms AB in the unit cell are different, changes in lattice vibration frequencies (in GaAs, for example, compared to Ge) arise in two ways. The second-neighbor A–A forces are different from the second-neighbor B–B forces, and the masses M_A and M_B are also different. The elastic constants depend only on $M_A + M_B$ (because the macroscopic strain field involves displacing only the masses of the unit cells) and for this reason chemical trends in elastic forces are easy to identify. In other cases the shift in frequencies from A = B (diamond-type) lattices to A $\neq$ B (zincblende) lattices is caused *both* by mass differences and by force differences. Usually the mass-induced shift is much larger than the force-induced one. The former becomes the first-order effect, the latter a second-order one.

Reliable interpretations of lattice vibrations in terms of certain kinds of interatomic forces are best obtained by studying chemical shifts within a family of crystals of like structure, as we saw in studying β/α vs. f_i. In order to examine chemical shifts in interatomic forces at short wavelengths, one should have data on a sequence of $A^N B^{8-N}$ crystals with A and B belonging to the same row of the periodic table, to minimize $(M_A - M_B)/(M_A + M_B)$. At present the cases $N = 3$ and $N = 4$ from the third row, i.e., Ge and GaAs, have been studied by neutron scattering, but the case $N = 2$, i.e., ZnSe, has not been measured. Thus until dispersion curves are available for ZnSe it does not appear likely that a definitive model of the dependence of interatomic forces of short wavelength on ionicity will be possible.

METALLIZATION IN GRAY Sn

In Chapter 1 we discussed the trend from covalent crystals like Si and Ge to metals like Pb which occurs within column IV of the periodic table as we go from light to heavy elements. This trend is called metallization, and the covalent-metallic transition takes place just at Sn, which at STP can be found either in the diamond structure (gray Sn), or in a distorted NaCl structure which is metallic (white Sn).

Even in gray Sn the incipient shear instability is indicated by the softness of the "ta" mode at the Brillouin zone edge, compared to other frequencies such as the Raman optic frequency at $q = 0$. This softness is shown for the entire diamond–Si–Ge–gray Sn family in Fig. 4.9. Another indication of the connection between this softness and the gray Sn–white Sn transition is obtained from pressure data. White Sn is about 20% more dense than gray Sn, so that application of pressure to a diamond-type crystal brings it closer to the covalent-metallic transition. In the absence of a phase transition, one would expect the lattice vibrational frequencies and the elastic constants of a crystal to increase under pressure. This means that the Grüneisen constants

$$\gamma_i = \partial \log\omega_i / \partial \log V = V \partial \omega_i / \omega_i \partial V \qquad (4.13)$$

should be positive, and this is normally found to be the case.

The Grüneisen constants are ordinarily difficult to measure at short wavelengths because of the limited resolving power of the neutron scattering method. However, in the special case of Ge it has proved possible to measure γ_i at the L point in tunneling experiments. The results are shown in Fig. 4.10. Note that γ_i is positive for the "la," "lo," and "to" modes, but that it is negative for the "ta" mode, in accord with our expectations. It appears that the gray Sn–white Sn first order phase transition is connected with the instability of gray Sn against short wavelength shear.

THERMAL EXPANSION

If the constants γ_i are all equal to a single Grüneisen constant, then β, the linear thermal expansion coefficient, is given by

$$\beta = \gamma(KC_v/3V), \qquad (4.14)$$

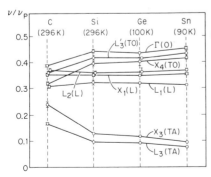

Fig 4.9. Lattice vibration frequencies of diamond-type crystals. All frequencies ν are reduced by the ion plasma frequency ν_p, which is the natural frequency of vibration of the ions in the lattice assuming no screening of the ion–ion Coulomb forces by valence electrons. (This is a convenient way of taking account of differences in lattice constant.) Notice the progressive softening of the transverse acoustic frequencies at the zone-boundary points X and L as we pass from C to Sn. [data and figure from D.L. Price and J. M. Rowe, *Solid State Commun.* **7**, 1433 (1969)].

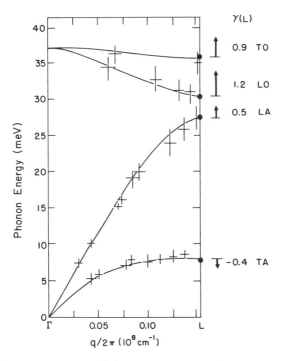

Fig. 4.10. Dispersion curves and Grüneisen constants for phonons at the L (zone-boundary) point in Ge [from H. Fritzsche, *in* "Tunneling Phenomena in Solids" (E. Burstein and S. Lundqvist eds.). Plenum Press, New York, 1969].

where K is the compressibility and C_v is the specific heat at constant volume. The values of γ_i shown in Fig. 4.10 explain the observation that Ge (and other tetrahedrally coordinated semiconductors) exhibits a negative value of β at low temperatures (around 50°K). At these temperatures the condition $kT \approx \hbar\omega$ for excited phonon modes yields excited phonons largely in the "ta" modes, where γ_i is negative. At room temperature, the average value of γ is positive, but still much smaller than is the case for most materials. Thus the thermal expansion coefficient for Ge at room temperature is about five times less than that of Zn.

VIBRATIONS OF IMPURITY ATOMS

When impurity atoms are present in a crystalline host, they modify the vibrational properties in a definite way. In addition to the band spectrum of lattice vibrations of the host crystal, there will be vibrational frequencies

ω_I associated mainly with localized motion of the impurity. If these frequencies lie within the band spectrum of the semiconductor host crystal, they will be difficult to separate from this band. However, if the mass M_I of the impurity atom is much smaller than M_h, that of the atoms of the host crystal, then one would expect that ω_I would be greater than ω_{max}, the maximum frequency of the band spectrum. In this case ω_I can be measured by infrared absorption. The oscillator strength of this absorption can be used to measure the effective charge associated with the impurity.

The condition $M_I < M_h$ is easily satisfied if the impurity atom belongs to the first period. Vibrations of Li, B and O atoms in silicon and N in GaP have been studied in this way. The Li and O atoms occupy interstitial sites, while B and N are substitutional. The vibrations of substitutional impurities are usually interpreted in terms of a simplified model in which the force constants between the substituted impurity and its nearest host neighbors are the same as host–host force constants. The frequency ω_I is then determined almost entirely from the band spectrum of the host and the relative mass defect $(M_h - M_I)/M_h$. If we let $\bar{\omega}_h$ be an average optical mode frequency of the host lattice, then for N in GaP one can calculate ω_I from the relation [6]

$$\omega_I^2 = \bar{\omega}_h^2 M_{Ga} M_P / [M_N (M_{Ga} + M_P)] \tag{4.15}$$

assuming that only the N atom vibrates in the local mode. Isotopic studies of B in Si have shown [7] that 90% of the vibrational mass of that local mode is the boron mass M_B. The value of $\hbar\omega_I$ calculated from $\hbar\bar{\omega}_h = 47$ meV is 58 meV, compared to the experimental value of 60 meV.

An interesting case of impurity vibrations is O in Si. Here the O enters the lattice in a position very close to the one it would have in SiO_2; that is, it lies nearly halfway between two Si atoms, but slightly displaced from the bond line, to form a bent triatomic molecule Si–O–Si. The vibrational frequencies in the crystal of the local modes associated with this oxygen impurity are close to those found in Si_2O groups in organic molecules [7, 8]. The O atom can tunnel from one off-center position to an adjacent one around the same bond line, giving rise to vibrational levels in the far infrared (~ 40 cm^{-1}), which are analogous to hindered rotations in molecular vibrational spectra.

SUMMARY

Short wavelength lattice vibrations have been studied in semiconductors generally by neutron scattering and specifically in indirect gap materials by tunneling and indirect optical transitions. From these studies some microscopic information has emerged, although less than what has been inferred from elastic constants (long wavelength vibrations). The additional information about interatomic forces concerns primarily the stability of tetrahedral bonds against shear, which is reduced as the gray Sn–white Sn transition is approached. Studies of optic mode splittings yield information concerning long-range Coulomb forces between partially charged ions.

REFERENCES

1. H. B. Rosenstock, *Phys. Rev.* **145,** 546 (1966).
2. M. Born and K. Huang, "Dynamical Theory of Crystal Lattices," pp. 89–100. Oxford Univ. Press (Clarendon), London and New York, 1954.
3. H. B. Callen, *Phys. Rev.* **76,** 1394 (1949); E. Burstein, *et al., J. Quant. Chem.* **15,** 759 (1967).
4. J. C. Phillips, *Phys. Rev.* **166,** 832 (1966).
5. R. M. Martin, *Phys. Rev.* **1B,** 4005 (1970).
6. D. G. Thomas and J. J. Hopfield, *Phys. Rev.* **150,** 680 (1966).
7. R. M. Chrenko *et al., Phys. Rev.* **138,** A1775 (1965).
8. D. R. Bosonworth *et al., Proc. Roy. Soc.* **A317,** 133 (1970).

5

Energy Bands

THE LANGUAGE OF BAND THEORY

The energy bands of solids, and particularly crystals with a large number of valence electrons per unit cell, appear quite complicated at first sight. This is especially true of semiconductors, for even the simplest ones have eight valence electrons per unit cell, compared for example to alkali metals, with only one conduction electron per unit cell. Fortunately much of this complexity disappears after one has learned what to expect. First, however, let us remind ourselves of the language of band theory.

The one feature of the quantum states of crystals which every student is expected to know is that the electronic wave functions have a special form imposed by the translational periodicity of the crystal. Each electronic wave function is assigned a crystal momentum $\mathbf{p} = \hbar\mathbf{k}$, where $\mathbf{k}$ is a wave vector, and a band index n. The wave function $\psi_{n\mathbf{k}}$ has the Bloch form

$$\psi_{n\mathbf{k}}(\mathbf{r}) = [\exp(i\mathbf{k}\cdot\mathbf{r})]u_{n\mathbf{k}}(\mathbf{r}), \tag{51.}$$

where $u_{n\mathbf{k}}(\mathbf{r})$ is a periodic function of $\mathbf{r}$ with the periodicity of the crystal lattice [1]. The energy of the state $\psi_{n\mathbf{k}}(\mathbf{r})$ is represented by $E_n(\mathbf{k})$.

Just as with lattice vibrations, the wave vector **k** ranges over reciprocal space, which is made up of Brillouin zones as discussed in Chapter 4. The first two Brillouin zones of a square lattice are shown in Fig. 5.1. The reciprocal lattice which is conjugate to a square lattice is itself square. The unit cell of a square lattice is square, so that this is the first Brillouin zone. The second Brillouin zone consists of the four shaded triangles shown in the figure. The area of the second Brillouin zone is equal to the area of the first zone, as one can see by folding back the four triangles into the first zone. The process of folding back becomes much more complicated for higher zones, and for other lattices, especially in three dimensions [2]. But in three dimensions there is also difficulty in displaying $E_n(\mathbf{k})$ in **k** space, especially outside the first Brillouin zone. Thus most often one sees $E_n(\mathbf{k})$ plotted only in the first Brillouin zone, the higher values of n (which correspond to higher zone numbers) being shown in the same figure as the $n = 1$ band.

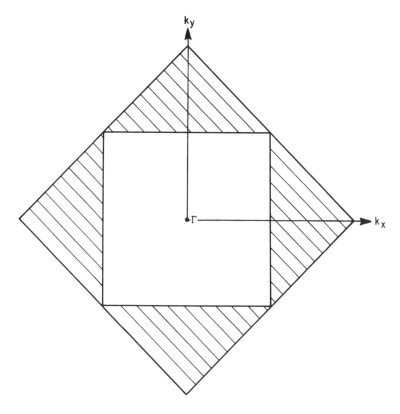

Fig. 5.1. The first and second Brillouin zones of the square lattice. The first Brillouin zone is square; the second zone consists of the four shaded triangles.

Although this procedure is convenient, some of the physics of $E_n(\mathbf{k})$ may sometimes be sacrificed this way. Next we discuss how to interpret energy bands of semiconductors which have been "folded back."

NEARLY FREE ELECTRON MODEL

It was observed more than three decades ago that the valence band width of semiconductors such as Si, as measured by soft X-ray emission, is close to the Fermi energy E_F of a free electron gas of density equal to that of the valence electrons, while in ionic crystals such as NaCl, the valence bands are narrow and resemble the atomic energy levels of the anions. The function $E_n(\mathbf{k})$ for a free electron gas is

$$E_n(\mathbf{k}) = \hbar^2(\mathbf{k} + \mathbf{G}_n)^2/2m, \tag{5.2}$$

where $\mathbf{G}_n$ is a reciprocal lattice vector and $\mathbf{k}$ is a wave vector in the first Brillouin zone. In one dimension $\mathbf{k}$ and $\mathbf{G}_n$ are necessarily parallel, and then (5.2) reduces to $\hbar^2 k_n^2/2m$, with $k_n = |\mathbf{k}| \pm |\mathbf{G}_n|$, a simple parabolic relation. In two and three dimensions $\mathbf{k}$ and $\mathbf{G}_n$ may not be parallel. Then the folded back energy bands will not look nearly parabolic, even when the energy gaps between bands, which are produced by the crystal potential, are fairly small.

VALENCE BANDS OF SILICON

To see how this works out in a particular case, consider $E_n(\mathbf{k})$ in Si. Here the first Brillouin zone is the face-centered cubic zone previously shown in Fig. 4.1. The smallest reciprocal lattice vectors, in units of $2\pi/a$, are $\mathbf{G}_0 = (0, 0, 0)$, $\mathbf{G}_1 = (1, 1, 1)$ and $\mathbf{G}_2 = (2, 0, 0)$ and their cubic permutations such as $(-1, 1, 1)$ and $(0, -2, 0)$. With eight valence electrons per unit cell, four valence bands are filled (with two electrons with opposing spins per cell). Looking along the (111) and (200) axes in reciprocal space (arbitrarily called Λ and Δ, respectively, in Fig. 4.1), the first energy band corresponds to $\mathbf{G} = \mathbf{G}_0 = 0$ in equation (5.2). The second energy band corresponds to the $\mathbf{G}_1$ (in the case of $\mathbf{k}$ along Λ) or $\mathbf{G}_2$ (in the case of $\mathbf{k}$ along Δ) which is *antiparallel* to $\mathbf{k}$. Since the value of $\mathbf{k}$ at the Brillouin zone boundary is $-\mathbf{G}_1/2$ (or $-\mathbf{G}_2/2$), these bands constitute parabolic continuations of band 1, if we neglect the small energy gap at the zone boundary. This is illustrated in Fig. 5.2.

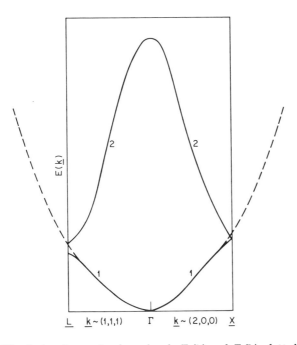

Fig. 5.2. The first and second valence bands $E_1(\mathbf{k})$ and $E_2(\mathbf{k})$ plotted against wave vector $\mathbf{k}$ in the first Brillouin zone for $\mathbf{k}$ along $(2,0,0)$ or $(1,1,1)$ directions. The second valence band is a nearly parabolic continuation of the first valence band which has been "folded back" into the first Brillouin zone by reflection in the point $\mathbf{X}$ or $\mathbf{L}$ for $\mathbf{k}$ in the $(2,0,0)$ or $(1,1,1)$ directions, respectively. Underlining in the figure corresponds to boldface letters elsewhere.

JONES ZONE

The third and fourth bands are less easily visualized. They involve either a $\mathbf{G}_1$ or a $\mathbf{G}_2$ which is neither parallel nor antiparallel to $\mathbf{k}$. The volume in reciprocal space containing the first n Brillouin zones (where $2n$ is the number of valence electrons per unit or primitive cell) is called the Jones zone [2]. For Si the Jones zone is the polyhedron shown in Fig. 5.3. Some of the symmetry points of the outer Si zones are shown also. For example, the centers of the faces of the Jones zones are labeled $\mathbf{X}$ in Fig. 5.3, and these points correspond to $\mathbf{k} = (2\pi/a)(0, 1, 1) = \mathbf{X}'$. According to Table 4.1, the coordinates of $\mathbf{X}$ in the *first* Brillouin zone are $(\pi/a)(2, 0, 0)$. This differs from $\mathbf{X}'$ by $\mathbf{G}_1 = (2\pi/a)(-1, 1, 1)$ and so $\mathbf{X}'$ in the outer zones does indeed differ from $\mathbf{X}$ by a reciprocal lattice vector. Moreover $\mathbf{G}_1$ is not parallel to $\mathbf{X}$, so we would expect that the energy bands in the third and

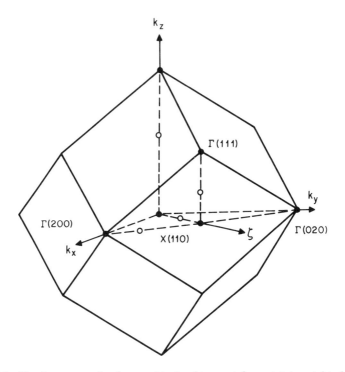

Fig. 5.3. The Jones zone for face-centered cubic crystals containing eight electrons per unit cell includes the first four Brillouin zones. The first face-centered cubic Brillouin zone was shown in Fig. 4.1. Points on the surface of the Jones zone which differ from points in the first zone by a reciprocal lattice vector are indicated with the symbols used to label points in Fig. 4.1.

higher zones near X' would not be obviously parabolic when plotted as a function of $\mathbf{k}$ along the symmetry line Δ connecting Γ to X.

SIMPLIFIED BANDS

One can simplify the energy bands in the third and fourth zones by assuming that $\mathbf{k}$ is approximately perpendicular to $\mathbf{G}$ and that $k^2 \ll G^2$. In this case $E_3(\mathbf{k}) \cong E_4(\mathbf{k}) \simeq \text{constant} \cong \hbar^2(X')^2/2m$. This is equivalent to saying that the E_3 and E_4 bands lie near the surface of the Jones zone, with $\mathbf{k}$ tangential to this surface, and that the energy levels of the surface region are nearly constant as a function of angle. This gives the energy levels $E_n(\mathbf{k})$, for $n = 1$ through 4, shown in the reduced zone in Fig. 5.4. The parabolic character of bands 1 and 2 arises from the fact that $\mathbf{k}$ is parallel to **G** for these bands, the nearly flat character of bands 3 and 4 from **k** being perpendicular to **G**.

ISOTROPIC MODEL

The picture shown in Fig. 5.4 can be idealized further in a way that will later help to clarify the relations between the band picture and chemical bonds. The polyhedron representing the Jones zone shown in Fig. 5.3 can be approximated by a sphere, and the anisotropic energy bands $E_n(\mathbf{k} + \mathbf{G})$ can be replaced by a single function $E(q)$, where $q = |\mathbf{k} + \mathbf{G}|$, that is, q is the crystal wave vector *not* reduced modulo **G**. In this isotropic model

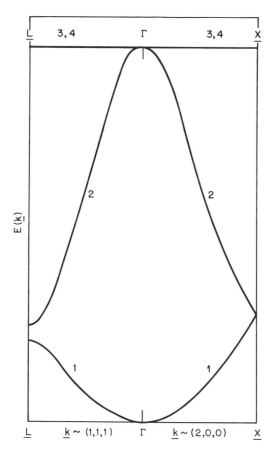

Fig. 5.4. Simplified valence bands ($n = 1$–4) for diamond-type crystals. Bands 1 and 2 are nearly parabolic; bands 3 and 4 are nearly flat.

the radius of the Jones zone is of course k_F, the Fermi wave number of a sphere in k space containing four electrons per atom. Because the sphere is a good approximation to the Jones polyhedron, k_F is about 10% larger than $|\mathbf{X'}|$. The important energy gap is the one between the occupied valence bands and the empty conduction bands, that is, between band 4 and band 5 in the first Brillouin zone. Other energy gaps (between band 1 and band 2, for example) are neglected. Previously we denoted this isotropic or average energy gap by E_g. Therefore $E(q)$ must be a function such that

$$E(k_F + \epsilon) - E(k_F - \epsilon) = E_g, \tag{5.3}$$

where $\epsilon \to 0$. Also for $q \ll k_F$ or $q \gg k_F$ we want $E(q)$ to approximate $\hbar^2 q^2/2m$, the free electron value.

SECULAR EQUATION

These requirements, if met in the simplest way, virtually determine the model energy function $E(q)$. To obtain the discontinuous behavior (5.3), we construct a model secular equation between two hypothetical plane wave basis states, $\exp(i\mathbf{q}\cdot\mathbf{r})$ and $\exp(i\mathbf{q'}\cdot\mathbf{r})$. Clearly $E(q) = E(-q)$, because of our isotropic approximation. Thus (5.3) involves $q = k_F + \epsilon$ and $q' = -k_F + \epsilon$ and $q' = q - 2k_F$. (Here $2k_F$ plays the role of a reciprocal lattice vector $\mathbf{G}$, and $\mathbf{G}$ is approximately equal to $2\mathbf{X'}$.) The model secular equation has the form

$$(H_{ij} - E\delta_{ij})\varphi_j = 0 \tag{5.4}$$

with matrix elements

$$H_{11} = \hbar^2 q^2/2m, \tag{5.5}$$

$$H_{22} = \hbar^2(q - 2k_F)^2/2m, \tag{5.6}$$

$$H_{12} = H_{21} = E_g/2. \tag{5.7}$$

The solution of equations (5.4)–(5.7) is given by

$$E(q) = (H_{11} + H_{22})/2 \pm \tfrac{1}{2}[(H_{11} - H_{22})^2 + E_g^2]^{1/2}, \tag{5.8}$$

where the $+$ sign corresponds to $q > k_F$, and the minus sign to $q < k_F$.

The qualitative behavior of (5.8) is shown in Fig. 5.5. Note that the valence energy bands are nearly flat within a shell of thickness $\delta = (E_g/4E_F)k_F$ of the surface. The volume of this shell is approximately $4\pi k_F^2\delta$, compared to the volume $4\pi k_F^3/3$ of the Fermi sphere. Thus the shell represents a fraction of the valence states given by $3E_g/4E_F$, or less than half of the valence states. This corresponds to bands 3 and 4 in Fig. 5.4 and

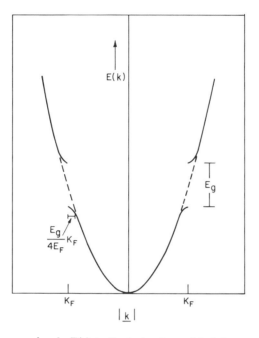

Fig. 5.5. The energy bands $E(q)$ in the isotropic model of the energy bands of diamond-type semiconductors. These bands are given explicitly by Eq. (5.8). They are shown as a function of $\mathbf{q} = \mathbf{k} + \mathbf{G}$, i.e., *not* "folded back."

shows that this model corresponds qualitatively to the situation discussed there.

Now that we have discussed the qualitative behavior of the valence bands, it is interesting to look at what the actual valence bands of Si or Ge look like. The valence bands of Si are shown in Fig. 5.6; those of Ge are quite similar. As one expects, bands 1 and 2 are nearly parabolic for **k** along the (111) and (200) symmetry directions, while bands 3 and 4 are nearly flat.

DIELECTRIC FUNCTION OF ISOTROPIC MODEL

The energy-band model shown in Fig. 5.5 can be extended to construct a model dielectric function $\epsilon(\omega)$. Let a long-wavelength electric field of frequency ω be incident on the crystal. The internal polarization at frequency ω is described by

$$\mathbf{D}(\omega) = \mathbf{E}(\omega) + 4\pi\mathbf{P}(\omega) = \epsilon(\omega)\mathbf{E}. \qquad (5.9)$$

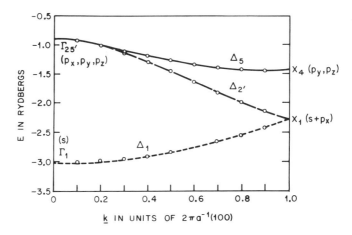

k IN UNITS OF $2\pi a^{-1}(100)$

Fig. 5.6. The valence bands of diamond plotted in the first Brillouin zone. These are accurate bands, calculated from the crystal Schrödinger equation by the methods discussed in Chapter 6. Their similarity to the simplified bands of Figs. 5.4 and 5.5 is emphasized in the text.

In general, of course, the dielectric function is a tensor, but in our isotropic model it is a simple scalar function. This is also true for all real crystals of cubic symmetry (such as Si and Ge) as well, so that in this respect the model is not oversimplified.

The dielectric function has both dispersive and absorptive parts. If the amplitude of the incident wave is proportional to $\exp(i\omega t)$, one can represent $\epsilon(\omega)$ by

$$\epsilon(\omega) = \epsilon_1(\omega) + i\epsilon_2(\omega), \qquad (5.10)$$

where $\epsilon_1(\omega)$ and $\epsilon_2(\omega)$ are real functions. The two functions are connected by the Kramers-Kronig relation [3]

$$\epsilon_1(\omega) = 1 + \pi^{-1} \int_{-\infty}^{\infty} \frac{\epsilon_2(\omega')\ d\omega'}{\omega' - \omega}. \qquad (5.11)$$

Because energy is conserved in an optical transition, valence electrons are excited from valence states of energy E_v to conduction band states of energy $E_c = E_v + \hbar\omega$. Such transitions are absorptive and so contribute to $\epsilon_2(\omega)$. Their contribution at other frequencies ω' to the dispersive part $\epsilon_1(\omega')$ of $\epsilon(\omega')$ is computed from (5.11).

Another quantity that must be conserved in an optical transition is crystal momentum. Because the wavelength of light at optical frequencies is several hundred lattice constants or more, the photon momentum is negligible compared to a typical electron momentum $\hbar\mathbf{k}$. This means for the initial valence band state $E_v(\mathbf{k})$ and the final conduction band state $E_c(\mathbf{k'})$ that for most practical purposes $\mathbf{k'} = \mathbf{k}$. On an $E(\mathbf{k})$ diagram plotted in the first Brillouin zone (like Fig. 5.6) such an optical transition can be represented by a vertical arrow.

At this point we discover something quite important. In the free electron limit ($E_g = 0$), the basis states are all pure states belonging to a single momentum, and optical transitions from $\mathbf{k}$ to $\mathbf{k'}$ are impossible. To have an optical transition, one must have states containing more than one plane wave. For instance, in our isotropic model $\mathbf{k}$ and $\mathbf{k'} = \mathbf{k}(1 - 2k_F)$ are mixed by $H_{12} = E_g/2$. This mixing is strong in the region near E_g where the energy bands are nearly flat. Thus in the isotropic model almost all the oscillator strength associated with excitation of the valence electrons is confined to states near k_F, that is, the $n = 3$ and 4 valence bands, and the $n = 5$ and 6 conduction bands. In fact when one evaluates $\epsilon_2(\omega)$ for the isotropic model one finds [4]

$$\epsilon_2(\omega) = 0, \qquad\qquad 0 \leq \hbar\omega \leq E_g, \quad (5.12)$$

$$\epsilon_2(\omega) = \frac{16}{3a_0k_F} \frac{E_F^2 E_g^2}{(\hbar\omega)^4} \frac{\hbar\omega}{[(\hbar\omega)^2 - E_g^2]^{1/2}}, \qquad E_g \leq \hbar\omega, \quad (5.13)$$

where $a_0 = \hbar^2/me^2$ is the Bohr radius. The relation (5.12) simply says that there is no real absorption until $\hbar\omega \geq E_g$. On the other hand, equation (5.13) shows that $\epsilon_2(\omega)$ is proportional to $(\hbar\omega - E_g)^{-1/2}$ as $\hbar\omega - E_g$ tends to zero. This shows that most of the absorption occurs for $\hbar\omega$ near E_g. Of course, in real crystals $\epsilon_2(\omega)$ never tends to infinity. This particular divergence occurs because of the isotropic character of the band model, and it disappears as soon as the real anisotropy of the energy bands is taken into account. However, the optical spectra of real semiconductors do exhibit behavior remarkably like (5.13), as we shall see later.

The Kramers–Kronig transform of (5.13) to give $\epsilon_1(\omega)$ is particularly simple for $\omega = 0$. Then one finds to a good approximation

$$\epsilon_1(0) = 1 + (\hbar\omega_p/E_g)^2(1 - E_g/4E_F). \quad (5.14)$$

Of course (5.14) is the relation that was made the basis of our analysis of chemical bonding in semiconductors in Chapters 1 and 2. The energy gap E_g was used there as a measure of the average difference in energy between bonding and antibonding states, represented as two narrow levels. The actual energy levels in the crystal, of course, are broadened into energy bands because of the periodicity of the crystal. In chemical language the bands arise because of interactions with second, third and more distant neighbors. However, it is the nearest-neighbor or bonding interactions that are most important.

To separate the nearest-neighbor interactions from the longer-ranged ones, it is necessary to average the energy bands and find the average energy difference between the mean energy of the bonding states and the mean energy of the antibonding states (see Fig. 1.1). Deforming the bond chemically or physically is analogous in quantum-mechanical language to exciting an electron optically, through an electric dipole transition, from the valence band to the conduction band. This is the reason we choose to determine the average energy gap E_g from the optical polarizability $\epsilon_1(0)$ by means of equation (5.14).

One may raise the mathematical question of whether this approach is unique (clearly it is not), and whether different approaches, using different averaging procedures, might yield different results. The first point to note is that the number of physically attractive averaging procedures is much smaller than the number of mathematically consistent ones. This is because the underlying philosophy of quantum mechanics is an operational one, and this philosophy states that physical observables [such as the polarizability $\epsilon_1(0)$] have special significance. Several other approaches to defining E_g have been proposed; so far, these have yielded results similar to those obtained from (5.14), but they appear to be less accurate in explaining chemical trends, especially in semiconductors.

In conclusion, then, the transformation of energy bands into chemical bonds will be carried out by using electronic polarizabilities and simplified band models to obtain quantities of chemical significance, such as E_h, C, and E_g. We now turn to a related problem, the properties of the valence bands and conduction bands nearest the minimum energy gap ΔE_{cv}.

IMPORTANT ANISOTROPIES

Although the anisotropy of the energy gap is not of primary importance in discussing the ionicity of the crystal bond, it is essential in understanding ΔE_{cv}, the *minimum* energy gap between valence and conduction bands. According to the definition of ΔE_{cv},

$$\Delta E_{cv} = E_c - E_v, \tag{5.15}$$

where E_c is the lowest energy in the conduction band and E_v is the highest energy in the valence band. Denote the values of crystal wave vector where the energies E_c and E_v occur by $\mathbf{k} = \mathbf{k}_c$ and $\mathbf{k} = \mathbf{k}_v$, respectively. Because of the nearly free-electron character of the energy bands $E_n(\mathbf{k})$, the locations of $\mathbf{k}_v$ and $\mathbf{k}_c$ tend to be much the same in all compounds having similar coordination configurations and the same number of valence electrons per atom. In the tetrahedrally coordinated $A^N B^{8-N}$ compounds, for example,

$\mathbf{k}_v$ nearly always occurs at $\mathbf{k} = 0$ (conventionally labeled Γ). The reason for this can be seen from Fig. 5.3, which shows that the corners of the Brillouin zone, which give the highest valence band energies because they are furthest from the origin in momentum space, correspond to the two reciprocal lattice vectors $\mathbf{G}_1 = (2\pi/a)\,(1, 1, 1)$ and $\mathbf{G}_2 = (2\pi/a)\,(2, 0, 0)$. Thus modulo these reciprocal lattice vectors the corners correspond to $\mathbf{k} = 0 = \Gamma$, and for strictly cubic symmetry the maxima of bands $n = 2$, 3 and 4 all occur at $\mathbf{k}_v = \Gamma$ with the same energy. The threefold orbital degeneracy is a consequence of cubic symmetry, which is removed by a uniaxial perturbation as in wurtzite (hexagonal) or chalcopyrite (tetragonal) crystals (Chapter 1).

CONDUCTION BANDS

The conduction bands are not shown in Fig. 5.6 because their behavior is complex and varies considerably from one crystal to another, for instance from Si to Ge. The lower conduction bands correspond to the nearly flat portions of the upper branch of Eq. (5.8) near the energy gap; that is, they correspond to $q \geq k_F$. Moreover the location of $\mathbf{k}_c$ varies from one compound to another. In compounds dominated by first or second row atoms $\mathbf{k}_c$ tends to fall at or near $\mathbf{k} = \mathbf{X}$; that is, the band edges tend to have axial $(1, 0, 0)$ symmetry. In compounds dominated by third, fourth or fifth row atoms $\mathbf{k}_c$ almost always is located at Γ; that is, its energy lies directly above $\mathbf{k}_v$ in the first Brillouin zone. The only exception to this rule is Ge, where $\mathbf{k}_c$ falls at $\mathbf{k} = \mathbf{L}$, which is the center of the $(1, 1, 1)$ face of the Brillouin zone. In compounds composed of atoms from the Si and Ge rows these three band edges fall within a few tenths of an eV of one another in all cases, but which lies lowest in a given compound appears to be accidental.

BAND-EDGE CURVATURES

Suppose the band-edge of interest is nondegenerate, and it occurs for $n = m$ and $\mathbf{k} = \mathbf{k}_0$. By definition at the band edge

$$(\partial E_m(\mathbf{k})/\partial \mathbf{k})_{\mathbf{k}=\mathbf{k}_0} = 0. \tag{5.16}$$

Let $E_m(\mathbf{k}_0) = E_0$, and expand $E_m(\mathbf{k})$ to second order in $(\mathbf{k} - \mathbf{k}_0)$:

$$E_m(\mathbf{k}) = E_0 + \sum_{i,j} a_{ij}(\mathbf{k} - \mathbf{k}_0)_i(\mathbf{k} - \mathbf{k}_0)_j, \tag{5.17}$$

where i and j label components of the vector $(\mathbf{k} - \mathbf{k}_0) = \delta\mathbf{k}$. The coefficients a_{ij} determine the curvatures of the energy surface $E = E_m(\mathbf{k})$ near $\mathbf{k} = \mathbf{k}_0$. The a_{ij} matrix can be diagonalized by referring it to principal axes, the correct choice of which is usually obvious from symmetry considerations. Referring to these axes by $i = 1, 2, 3$ we simplify (5.17) to

$$E_m(\mathbf{k}) = E_0 + \sum_i \frac{(\hbar^2)}{(2m_i{}^*)} (\delta\mathbf{k}_i)^2, \qquad (5.18)$$

where $\hbar^2/2m_i{}^*$ represents the principal values of the a_{ij} matrix. One now says that particles near this band edge have an effective mass $m^* = m_i{}^*$ if they are moving in the ith direction. (This expression will be justified in detail by wave-packet analysis in Chapter 9.) The values of $m_i{}^*$ can differ greatly from those of the free-electron mass m.

PERTURBATION THEORY

If the energy-band structure is sufficiently well known, the values of the effective masses $m_i{}^*$ can be calculated from perturbation theory. Although the resulting formulas appear to be quite complicated, the basic idea is simple. Starting from the Bloch form (5.1), we evaluate $u_{mk}(\mathbf{r})$ in terms of $u_{nk_0}(\mathbf{r})$, and simultaneously express $E_m(\mathbf{k})$ in terms of $E_n(\mathbf{k}_0)$. We compare the wave equation $H\psi = E\psi$ at $\mathbf{k}$ with the same equation at $\mathbf{k}_0$. Because $\mathbf{k} - \mathbf{k}_0$ is small, the difference between the two equations is small, and this difference can be treated by perturbation theory.

The Hamiltonian H of the crystal consists of two parts, the kinetic energy T of the electrons and their potential energy V. The kinetic energy operator T is given by

$$T = \mathbf{p}^2/2m = (-\hbar^2/2m)(\partial^2/\partial x^2 + \partial^2/\partial y^2 + \partial^2/\partial z^2). \qquad (5.19)$$

By explicit differentiation

$$(\partial/\partial x)\exp(i\mathbf{k}\cdot\mathbf{r})u_k(\mathbf{r}) = \exp(i\mathbf{k}\cdot\mathbf{r})[(\partial/\partial x) + ik_x]u_k(\mathbf{r}), \qquad (5.20)$$

which can be written in vector form using $\mathbf{p} = -i\hbar\nabla$ as

$$\mathbf{p}\psi_k = \exp(i\mathbf{k}\cdot\mathbf{r})(\mathbf{p} + \hbar\mathbf{k})u_k \qquad (5.21)$$

and in a similar way

$$\mathbf{p}^2\psi_k = \exp(i\mathbf{k}\cdot\mathbf{r})(\mathbf{p}^2 + 2\hbar\mathbf{k}\cdot\mathbf{p} + \hbar^2k^2)u_k. \qquad (5.22)$$

The Schrödinger equation $H\psi = E\psi$ becomes

$$(T + V)\psi_{mk} = \exp(i\mathbf{k}\cdot\mathbf{r})(\mathbf{p}^2/2m + \hbar^2k^2/2m + V + \hbar\mathbf{k}\cdot\mathbf{p}/m)\mathbf{u}_{mk}$$

$$= \exp(i\mathbf{k}\cdot\mathbf{r})E_m(\mathbf{k})u_{mk}. \qquad (5.23)$$

Canceling the common prefactors and defining $E_m'(k) = E_m(\mathbf{k}) - \hbar^2 k^2/2m$, we can rewrite equation (5.23) as

$$(T + V + \hbar\mathbf{k}\cdot\mathbf{p}/m)u_{mk} = E_m'(\mathbf{k})u_{mk}. \tag{5.24}$$

When the same derivation is repeated for $\mathbf{k} = \mathbf{k}_0$, we see that the parentheses on the left side is changed only by the term $\hbar(\mathbf{k} - \mathbf{k}_0)\cdot\mathbf{p}/m = \hbar\delta\mathbf{k}\cdot\mathbf{p}/m$. For this reason this approach is called $\mathbf{k}\cdot\mathbf{p}$ perturbation theory.

The standard formulas of perturbation theory can now be used to calculate the coefficients a_{ij} or $\hbar^2/2m_{ij}^*$. It is helpful to see these explicitly for the simple case of a nondegenerate band edge: (labeled $n = 0$),

$$\frac{m}{m_{ij}^*} = \delta_{ij} + \frac{2}{m}\sum_{n\neq 0}\frac{\langle u_0 \mid \mathbf{p}_i \mid u_n\rangle\langle u_n \mid \mathbf{p}_j \mid u_0\rangle}{E_0 - E_n}. \tag{5.25}$$

When the band edges are degenerate, one must use degenerate perturbation theory, but otherwise little is changed.

One of the reasons that (5.25) is useful is that it has been found in many cases that there are only one or two large terms in the sum on the right-hand side. These come from bands close in energy to the initial band, so that $E_0 - E_n$ is small, and it often can be measured optically. The terms in the numerator have the same form as optical oscillator strengths, and they are obtainable either from theoretical calculations or from experiment, as they are found to vary slowly from one compound to another. In a few cases, such as the valence band edges in Si and Ge, all the important terms in (5.25) are known, and these can be used to estimate values in less studied materials.

SPECIAL CASES

The valence bands in the neighborhood of $\mathbf{k} = 0 = \Gamma$ are shown in Fig. 5.7 for almost all diamond and zincblende-type semiconductors. Because these crystals are cubic, bands 2, 3, and 4 are orbitally degenerate at Γ in a way that is analogous to the orbital degeneracy of p^3 (x, y, and z) states in atoms. When the electron spin is included, one has $3 \times 2 = 6$ states altogether, which are split into $J = \frac{3}{2}$ (fourfold degenerate) and $J = \frac{1}{2}$ (twofold degenerate) states at $\mathbf{k} = 0$. The splitting is caused by the spin–orbit interaction $\lambda\mathbf{L}\cdot\mathbf{S}$, where λ is the spin–orbit splitting parameter. Its value in the crystal is found to be about 50% greater than in the free atom. (A longer discussion of the meaning of λ is given in Chapter 7.) Because the spin–orbit interaction is concentrated mainly in the atomic cores, this enhancement reflects the compression of the valence states on passing from the atomic to the crystalline state.

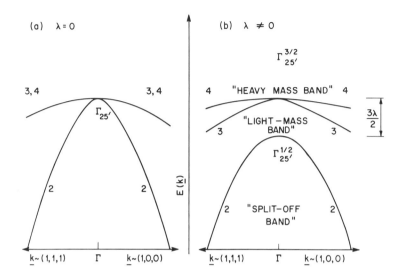

Fig. 5.7. Sketch of the valence bands of diamond and zincblende-type semiconductors on a greatly expanded scale near the top of the valence band at $\mathbf{k} = \Gamma = 0$. In (a) the bands are shown without the effect of spin–orbit interactions. In (b) spin–orbit splitting is included and the usual terminology is indicated.

In Fig. 5.7 we see three bands, each of which is twofold or spin-degenerate in the absence of a magnetic field. The lowest band is the $J = \frac{1}{2}$ or split-off band, and the two higher $J = \frac{3}{2}$ bands are called the light-mass and heavy-mass band, the latter energy being nearly independent of $\mathbf{k}$. These two bands can also be labelled by $m_J = \frac{3}{2}$ or $\frac{1}{2}$, the azimuthal quantum number of $\mathbf{J}$ relative to the $\mathbf{k}$ axis. Values of the effective mass parameters for these bands are known in many crystals [5].

For the conduction bands the situation is simpler because each band edge is nondegenerate. At Γ the conduction band is very much like the mirror image of the split-off valence band in Fig. 5.7. In a few crystals (gray Sn, HgS, HgSe, and HgTe) the conduction band actually touches the valence band at $\mathbf{k} = 0$, as shown in Fig. 5.8. This accident gives rise to unusual properties, including rather unstable crystal structures, in these materials.

When the conduction edge of band 5 is at $\mathbf{L}$ or along a $(1, 0, 0)$ axis near $\mathbf{X}$, the masses have uniaxial or cylindrical symmetry and the band-edge energies are given by

$$E_m(\mathbf{k}) = E_m(\mathbf{k}_0) + \hbar^2 \delta k_{||}{}^2/2m_{||}{}^* + \hbar^2 \delta k_{\perp}{}^2/2m_{\perp}{}^*, \qquad (5.26)$$

where $||$ and perpendicular refer to the appropriate symmetry axis, $(1, 1, 1)$ near $\mathbf{L}$, $(1, 0, 0)$ near $\mathbf{X}$. In all cases $m_{||}{}^*$ is close to m, while $m_{\perp}{}^*$ is about

0.1 to 0.3m. One can show from (5.25) that this is because the only important nearby bands are bands 3 and 4 in the valence band, and that these affect only $m_\perp^*$. The cylindrical energy surfaces of **L** and **X** band edges are shown in Fig. 5.9.

ATOMIC ORBITALS

So far in this chapter the emphasis has been on using the nearly free electron model to describe energy bands qualitatively. One can also represent the crystal wave function ψ as a linear combination of atomic s, p, and d orbitals centered on each atom. This is the approach used to describe wave functions in molecules. It is less satisfactory in semiconductors, for reasons to be examined in detail in the next chapter. However, it does provide a useful description of bands 3 and 4, and of the conduction bands

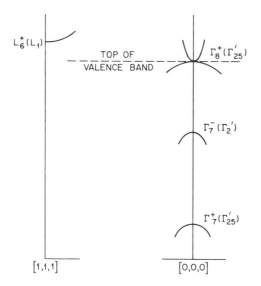

Fig. 5.8. A sketch of the energy bands of gray Sn. In this case it is essential to include the effects of spin–orbit splitting, for the spin–orbit splitting energy λ is of order 1 eV, compared to an orbital splitting between Γ_{25}' and Γ_2, of order 0.11 eV. The effect of spin–orbit splitting has been shown in Fig. 5.7, where we saw that the threefold degenerate (p-like) level Γ_{25}, splits into a $J = \frac{3}{2}$ and $J = \frac{1}{2}$ level. These levels can also be labeled Γ_8^+ and Γ_7^+, respectively, while when spin is included the label of the nondegenerate level becomes Γ_7^-, as shown here. The sketch shows that Γ_2' (or Γ_7^-) falls between the spin–orbit split components of Γ_{25},. As a result, in pure gray Sn the top of the valence band and the bottom of the conduction band coincide at Γ_8^+, and we have a zero-gap semiconductor.

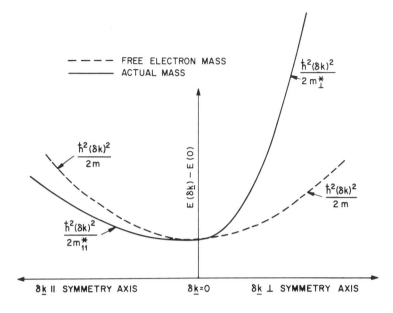

Fig. 5.9. Band curvatures for conduction band edges at or near **L** and **X**. The principal directions are those parallel and perpendicular to the ΓL or ΓX symmetry axes, respectively. Also shown dotted are the band curvatures for the effective masses equal to the free electron value. The quantity δk is the displacement of **k** from the band edge.

at Γ. It is included here for the reader's convenience, and the results are summarized in Fig. 5.10.

In atomic orbital language the wave functions for bands 3 and 4 are constructed from p-atomic orbitals oriented perpendicular to the wave vector **k**. Thus if **k** is along a $(0, 0, 1)$ axis the corresponding atomic wave functions for bands 3 and 4 are p_x and p_y wave functions centered on each atom with their amplitudes in phase (bonding state). Because of cubic symmetry the energies of the p_x and p_y states are equal, and bands 3 and 4 are degenerate (the bands stick together) for **k** along $(1, 0, 0)$. In partially ionic AB compounds, the weight given the wave function on atom A is different from that for atom B, so that

$$\psi \sim \varphi_A + \gamma \varphi_B, \qquad (5.27)$$

where $\gamma > 1$ and φ_A and φ_B are normalized atomic p states.

In the conduction band at $\mathbf{k} = 0$ there are two kinds of states of importance, with the antibonding form

$$\psi \sim -\gamma \varphi_A + \varphi_B, \qquad (5.28)$$

with $\gamma = 1$ when A = B. The nondegenerate state is formed from s orbitals

φ_A and φ_B, the orbitally threefold degenerate state from p orbitals $\varphi_A{}^x$, $\varphi_B{}^x$, or $\varphi_A{}^y$, $\varphi_B{}^y$, etc. At locations in the conduction band other than $\mathbf{k} = 0$ the atomic orbital description may be quite misleading, and it is desirable to discuss behavior elsewhere using more sophisticated models.

The atomic orbital approach is especially useful for describing valence bands near Γ. The threefold orbital degeneracy at Γ corresponds to p_x, p_y and p_z atomic bonding orbitals. A uniaxial perturbation, such as a potential proportional to $3z^2 - r^2$, splits the p_z band from the p_x and p_y bands. In this context atomic orbitals provide a useful simplified picture, but real understanding requires a complete description of all valence and conduction bands near the energy gap. This is obtainable from the pseudopotentials discussed in the following chapter.

SPECIFIC BAND STRUCTURES

Although the energy bands $E_n(\mathbf{k})$ of many $A^N B^{8-N}$ semiconductors can be considered approximately as "folded-back" versions of the $E(q)$ curve

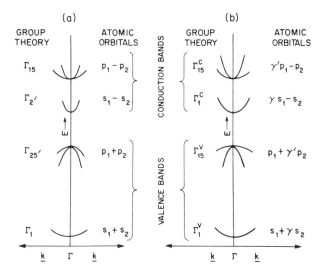

Fig. 5.10. Valence and conduction bands near Γ ($\mathbf{k} = 0$) according to the tight-binding prescription. In (a) the bands are labeled for diamond-type crystals, in (b) for zincblende type crystals. Spin–orbit splitting effects are omitted. The atomic orbitals considered are either s-type or p-type, the subscripts referring to atoms 1 and 2 in the unit cell. Also shown are the conventional group theory labels for the same states. The latter are discussed in a number of places, for example, G. F. Koster, *Solid State Phys.* 5 (1959).

shown in Fig. 5.5, in practice we are usually interested only in the bands within at most 2 or 3 eV of the Fermi energy. These bands are approximately "flat" in the simplified folded-back model of Fig. 5.4. However, the variations of these bands from one material to another are sufficiently great (especially in the conduction bands) that separate discussions of the two highest valence bands (numbered 3 and 4) and two lowest conduction bands (numbered 5 and 6) are appropriate for different crystals. The following cases also serve to illustrate the general principles discussed earlier in the chapter.

DIAMOND AND SILICON

Diamond is the hardest known material, with a very high Debye temperature (above 2000°K). Silicon is also very strong, and as we saw in Table 2.4, its cohesive energy, when expressed in reduced units, is unusually large relative to other diamond-type crystals. The energy bands of silicon are shown near the energy gap in Fig. 5.11. Both the valence and conduction bands of the two crystals are quite similar. The average energy gap $E_g = E_h$ is, of course, much greater in diamond than in silicon (13.6 eV compared to 4.8 eV), because the bond length d is smaller, and E_h scales like $d^{-2.5}$. The widths of the valence and conduction bands vary less rapidly, however, with d. Thus, although the maxima of the valence band and minima of the conduction bands correspond in both crystals to $\mathbf{k}_v = 0$, $\mathbf{k}_c \cong 0.8\mathbf{X}$, the minimum energy gap ΔE_{cv} is 5.7 eV in diamond and only 1.1 eV in silicon, as shown in the figures. In both crystals there are no other low-lying conduction band edges, and the lowest conduction band state at Γ is labeled Γ_{15} (antibonding p states, according to atomic orbital language, Fig. 5.9)

GERMANIUM AND GALLIUM ARSENIDE

The energy bands of Ge near the energy gap are shown in Fig. 5.12. Although the lattice constant of Ge is only 4% larger than that of Si, there are important shifts in conduction band energies at $\mathbf{k} = \Gamma = 0$ and $\mathbf{k} = \mathbf{L} = (\pi/a)\,(1, 1, 1)$. These levels are denoted by $\Gamma_{2'}$ (antibonding s, according to Fig. 5.9) and L_1, respectively. On going from Si to Ge, $\Gamma_{2'}$ drops by 3.1 eV and L_1 drops by 1.3 eV, or almost half as much. The valence bands and other conduction bands change little on going from Si to Ge.

The changes in $\Gamma_{2'}$ and L_1 arise from their antibonding s-character. In the following chapter we will see that this antibonding s-character is only half as well defined for L_1 as it is for $\Gamma_{2'}$. The antibonding s levels drop in

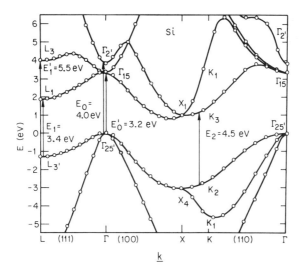

Fig. 5.11. Pseudopotential calculations of the energy bands $E_n(\mathbf{k})$ of silicon near the energy gap. Optical transitions between the valence and conduction bands are indicated by arrows. The labeling of the transitions is conventional, but the interband energies quoted are the ones derived from experiment, as discussed further in Chapter 7. The open circles are simply the calculated values of $E_n(\mathbf{k})$, while the solid lines are curves drawn through the calculated points. In these calculations the spin–orbit interaction is omitted for simplicity, and the symmetries of the wave functions at symmetry points or along symmetry lines are indicated in some cases. See [6].

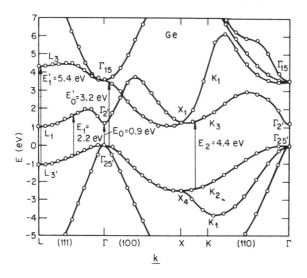

Fig. 5.12. Pseudopotential calculations of the energy bands $E_n(\mathbf{k})$ of germanium.

energy compared to the top of the valence band (bonding p) because of the trends in atomic s energies compared to atomic p energies (see Fig. 1.7).

In both diamond and Si $\Gamma_{2'}$ (antibonding s) is at about the same energy as Γ_{15} (antibonding p), i.e., at about an average conduction band energy, which in Si is 4 eV above E_v, whereas L_1 is only 2 eV above E_v. On the other hand, in Ge L_1 manages to fall slightly below $\Gamma_{2'}$ and the other conduction band edge near $\mathbf{X}$. Thus the bottom of the conduction band in Ge is at $\mathbf{L}$, a situation unique to this crystal. In other cases either L_1 and $\Gamma_{2'}$ are too high, and the conduction band minimum falls at or near $\mathbf{X}$ (Si, GaP) or L_1 and $\Gamma_{2'}$ have dropped further, with $\Gamma_{2'}$ overtaking L_1 because it drops faster. In Ge and GaAs all three conduction band edges have energies that are equal to within a few tenths of an electron volt. The situation in GaAs is shown in Fig. 5.13. Here we see that $\Gamma_{2'}$, now called Γ_{1c} (Fig. 5.7) is lowest, so that $\mathbf{k}_c = \mathbf{k}_v$, making GaAs a *direct-gap* semiconductor, in contrast to diamond, Si and Ge, where $\mathbf{k}_c \neq \mathbf{k}_v$ (indirect gap).

INDIUM ANTIMONIDE AND ARSENIDE

The compounds InSb and InAs have band structures similar to that of GaAs, except that the level Γ_{1c} is very close to the highest valence band state $^{3/2}\Gamma_{15}^v$. The energy bands of InAs are shown in Fig. 5.14. Because Γ_{1c} is so close to $^{3/2}\Gamma_{15}^v$, the absorption edges at room temperature are at 0.18 eV

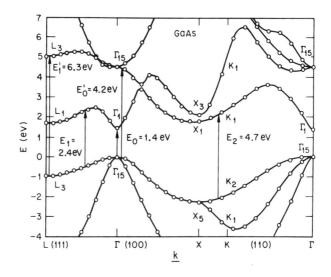

Fig. 5.13. Pseudopotential calculations of the energy bands $E_n(\mathbf{k})$ of gallium arsenide.

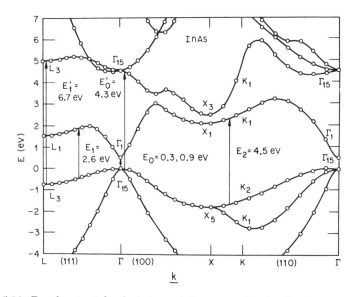

Fig. 5.14. Pseudopotential calculations of the energy bands of indium arsenide.

for InSb and 0.33 eV for InAs. This means that these semiconductors are transparent (apart from absorption due to thermally excited carriers) only in the far infrared. For this reason one may call them small gap semiconductors. In the near infrared there is strong intrinsic absorption.

GRAY TIN AND THE MERCURY CHALCOGENIDES

The element Sn is found at STP either in the white (metallic) form or in the gray (diamond) structure. Both gray Sn and HgX (X = S, Se or Te) exhibit energy bands in which the conduction band level $\Gamma_{2'}$ or Γ_{1c} lies below the $J = \frac{3}{2}$ valence band maximum. As a result, the energy levels at Γ of gray Sn are rearranged as shown in Fig. 5.8. In order to give a consistent discussion of this rearrangement, it is necessary to label the levels according to both orbital and spin symmetry, as indicated. The top of the valence band and the bottom of the conduction band touch at $\mathbf{k} = 0$, as shown in the figure. Thus these crystals can be described as zero-gap semiconductors ($\Delta E_{cv} = 0$, but $E_g > 0$).

EFFECTIVE MASS PARAMETERS

The curvatures or effective masses at valence and conduction band edges of $E_n(\mathbf{k})$ are known for several crystals from a number of experiments to be described in later chapters. All the experiments are most accurate at low temperatures in pure crystals. However, values appropriate to room temperature and to other compounds, which have so far not been prepared in purified form, can be estimated from the few accurate experimental values. This is done by using equations such as (5.25) and exploiting values of the energy denominators $E_0 - E_n$ as obtained from optical experiments. The optical oscillator strengths in the numerator of (5.25) vary slowly with changes in temperature or chemical composition and hence can be obtained by interpolation and extrapolation from the accurately known experimental values. Here we discuss the results obtained in this way for several conduction band edges which are orbitally nondegenerate. The more complex case of the effective mass parameters of the orbitally degenerate valence bands is discussed elsewhere [5].

For most compound semiconductors the conduction band edge occurs at Γ ($\mathbf{k} = 0$) and is labeled Γ_1 (in diamond-type crystals $\Gamma_{2'}$). This edge has s-like symmetry, is nondegenerate, and is described by a single effective mass parameter $m_c{}^*$. Qualitatively speaking, $m_c{}^*$ is inversely proportional to E_0, the energy gap between the top of the valence band (labeled $\Gamma_{15}^{3/2}$ in Fig. 5.7) and Γ_1. Accurate values of E_0 and $m_c{}^*/m$, where m is the free electron mass, estimated for $T = 0°\mathrm{K}$, are listed for a number of compound semiconductors in Table 5.1. Note that in compounds with the gray Sn band structure $m_c{}^*$ is negative (see Fig. 5.8). The smallest positive masses are found in InSb and InAs ($m_c{}^*/m = 0.014$ and 0.023, respectively).

In diamond, Si, Ge, and GaP the lowest conduction band edges are not at Γ, but they are nondegenerate and can be characterized by longitudinal and transverse effective masses $m_l{}^*/m$ and $m_t{}^*/m$, the values of which are listed in Table 5.2. Also listed are the multiplicity (number of symmetrically equivalent edges) and positions of the band edges.

THE PbS FAMILY

The energy bands discussed so far are those of $A^N B^{8-N}$ compounds crystallizing in tetrahedrally coordinated structures. However, we mentioned in Chapter 1 that because of the lowering of s energy levels relative to p levels as we go down any column of the periodic table, eventually the tetrahedrally coordinated structures become unstable, and usually compounds composed of such heavy elements are either unstable or are metallic. The simplest example of this behavior is the column IV sequence itself, diamond, Si, Ge, gray Sn, white Sn, and Pb. Here the first crystal may be called either a semiconductor or an insulator, according to one's taste. Silicon and germanium are the classic semiconductors. Gray Sn, as we just

TABLE 5.1

Effective Conduction-Band Masses m_c^* at Γ for a Number of
Semiconductors at $T = 0°K^a$

Crystal	E_0 (eV)	m_c^*/m (experiment)	m_c^*/m (theory)[a]	$P^2/2m$ (eV)
Ge	0.89	0.038	0.038	22.5
Gray Sn	−0.41	−0.058	−0.058	—
GaP	2.87	—	0.17	—
InP	1.42	0.077	0.080	17
AlAs	3.05	—	0.22	—
GaAs	1.52	0.066	0.067	21.5
InAs	0.42	0.023	0.023	17.8
AlSb	2.30	—	0.18	—
GaSb	0.81	0.045	0.045	17.2
InSb	0.24	0.014	0.014	16.9
ZnS	3.80	—	0.28	—
HgS	−0.15	—	−0.006	—
ZnSe	2.82	0.16	0.14	14.8
HgSe	−0.24	—	−0.042	—
ZnTe	2.39	—	0.18	—
CdTe	1.60	0.096	0.096	15.0
HgTe	−0.30	—	−0.031	—

[a] The theoretical values are partly fitted to the experimental ones and partly interpolated from the experimental values of E_0 using sum rules like Eq. (5.25). The theoretical values correct for small chemical trends in $P^2/2m = E_0 \, (m/m_c^* - 1)$. As shown in the last column, this quantity, when calculated from experimental values, is nearly constant. See [5] for a discussion of chemical trends in the interband dipole moments P.

TABLE 5.2

Effective Conduction Band Masses (for Band Edges not at Γ)
for Several Semiconductors[a]

Crystal	Multiplicity	Position	m_l^*/m	m_t^*/m
Si	6	0.85 ΓX	0.92	0.19
GaP	3	X	1.5	0.18
Ge	4	L	1.59	0.082

[a] The values quoted are obtained from cyclotron resonance experiments and from infrared optical spectra in GaP [from A. Onton and R. C. Taylor, *Phys. Rev. B* 1, 2587 (1970)].

saw, is a zero-gap semiconductor; it may also be called an ideal semimetal. White Sn is a good metal, but it has a complex structure, with each atom approximately sixfold coordinated, but covalently distorted away from a simple cubic structure. Finally Pb has a normal, close-packed metallic structure.

The PbS family includes as well PbSe and PbTe, SnSe and SnTe, GeTe and the semimetals As, Sb, and Bi. All the compounds are semiconductors, with the NaCl crystal structure, or nearly so (see Chapter 8 for further discussion of the structures). In the case of PbTe, the occupied energy levels are simply $s^2 p_+^3$, in the notation of Chapter 1 and Table 1.3. The situation for the remaining compounds is not so clear, for reasons we can discuss qualitatively.

The central feature of compounds of the PbS family which makes them of practical interest is that they all have unusually small direct-energy gaps which lie in the infrared. In eV in the Pb compounds these gaps are: PbS, 0.28; PbSe, 0.16; and PbTe, 0.19. In SnTe the gap is 0.25 eV, close to the value in PbTe. The space lattice for these $A^N B^{10-N}$ compounds is face-centered cubic, just as it is for the $A^N B^{8-N}$ compounds. However, the small energy gaps in the latter tend to occur at Γ (see Fig. 5.14 for InAs), while in the Pb salts the small gaps occur at the point L, which is the center of the (111) face of the Brillouin zone.

The general energy level scheme at the point L is shown in Fig. 5.15. Just as was the case for gray Sn, the spin–orbit splittings are as large or larger than the orbital splittings, so the levels are shown without spin-orbit splittings and with them. The energy gap in this scheme is between the antibonding s state (labeled L_1 or L_6^+) and an antibonding p state (labeled L_2, or L_6^-).

Optical studies have shown, however, that the levels which constitute the valence and conduction band edges in $Pb_x Sn_{1-x} Te$ alloys cross between $x = 0.4$ and $x = 0.6$. This means that the simple ordering of energy levels shown in Fig. 5.15 cannot be correct for both compounds; probably it is correct for neither. An ordering of energy levels based on pseudopotential calculations is shown in Fig. 5.16. In this ordering the valence band maximum in PbS, PbSe, and PbTe is the antibonding s state L_{1+}^{6+} (the single and double group notations are both exhibited), but in SnTe it is L_{3-}^{6-}, a state of antibonding p orbital symmetry. There are also differences in the symmetries of the conduction band minima, which appear to be in agreement with measured effective masses, g factors, and the temperature dependences of the band gaps.

While we are on the subject of crossing of conduction and valence band edges in small gap semiconductors, note that crossing at Γ can be achieved in $Hg_x Cd_{1-x} X$ alloys (X = S, Se or Te). In the $Hg_x Cd_{1-x} Te$ alloys the

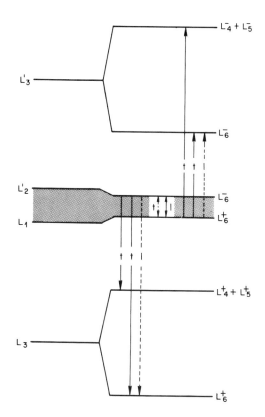

Fig. 5.15. A simple schematic energy-level diagram for electron energy levels at **L** for $A^N B^{10-N}$ semiconductors. On the left are the energy levels without inclusion of spin–orbit effects, with the energy gap between valence and conduction bands shaded. On the right spin–orbit effects have been added, with the assumption (not actually valid) that these are small compared to separations of the levels on the left.

crossing occurs near $x = 0.84$ and $T = 60°K$. Below this temperature one has the normal (InAs or InSb) ordering of levels, above it the "inverted" or gray Sn ordering [7]. Thus the level reversal can be achieved at fixed temperature by varying composition or, for some fixed compositions, by varying temperature.

The energy bands of $SnTe(A^N B^{10-N})$ exhibit near the point $\mathbf{k} = \mathbf{L}$ an unusual structure which arises because of energy levels just discussed. The maxima of the valence band and minima of the conduction band are not exactly at $\mathbf{k}_\alpha^0 = \mathbf{L}$, but are instead "off-center", as shown in Fig. 5.17. Because of spin–orbit effects, this behavior should exist in principle in sphalerite crystals ($A^N B^{8-N}$) such as GaAs or InSb for the top of the valence band, but the energy displacement (of order 10^{-4} eV, compared to 0.1 eV in

SnTe) has proved too small to be detected experimentally. Effects of the kind shown in Fig. 5.17 are also observed in $A^N B^{9-N}$ sandwich semiconductors, such as GaSe, with energy displacements again of order 0.1 eV. Such effects make it difficult to expand $E(\mathbf{k})$ about the band edge $\mathbf{k} = \mathbf{k}_\alpha$, because there are many $\mathbf{k}_\alpha$ clustered near the point $\mathbf{k}_\alpha{}^0$. This in turn means that the utility of wavepacket or effective mass theories (cf. Chapter 10) is much reduced in such cases. Fortunately, most semiconductors belong to the $A^N B^{8-N}$ family, and so effects of the kind shown in Fig. 5.17 are at present to be regarded as curiosities.

SUMMARY

The energy bands of elemental semiconductors resemble those of a free electron gas in two respects. First the gaps between bands are small compared to the overall valence bandwidth. Second when the energy bands are

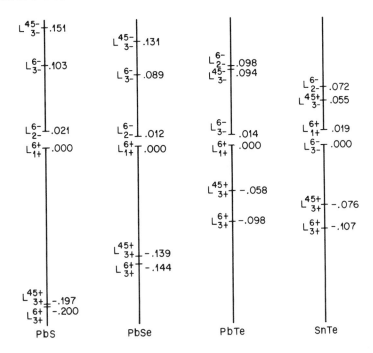

Fig. 5.16. Energy levels at **L** in PbS and related semiconductors; the highest valence level has been chosen as the zero of energy in each case, and this level lies near the middle of each scale. Calculations are based on semi-empirical pseudopotentials. [from R. L. Bernick and L. Kleinman, *Solid State Commun.* **8**, 569 (1970)]. Note the reversal of conduction and valence band edges between PbTe and SnTe.

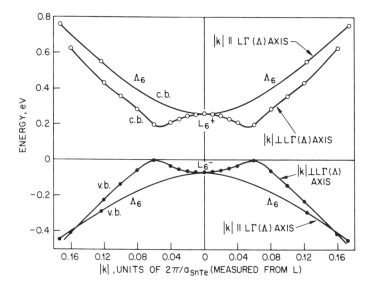

Fig. 5.17. Energy bands near the energy gap and near the point L in SnTe [from Y. W. Tung and M. L. Cohen, *Phys. Lett.* **29A**, 236 (1969)].

plotted as a single-valued function of $q = k + G$, they follow closely the free electron energy–momentum relation $E(q) = \hbar^2 q^2/2m$. This relation is obscured by the necessity of "folding back" the bands of outer Brillouin zones into the first zone for convenience of representing the three-dimensional variable q. Near the energy gap, however, important differences from the free electron model are found. These are described at band edges by using effective masses, which can be calculated by second-order perturbation theory. Surveys of many semiconductor band structures have been carried out with pseudopotentials [6].

REFERENCES

1. C. Kittel, "Introduction to Solid State Physics," p. 259. Wiley, New York, 1966.
2. H. Jones, "The Theory of Brillouin Zones," p. 195. North-Holland Publ., Amsterdam, 1960.
3. L. D. Landau and E. M. Lifshitz, "Electrodynamics of Continuous Media," p. 259. Pergamon, Oxford, 1960.
4. J. C. Phillips, *Rev. Mod. Phys.* **42**, 346 (1970).
5. P. Lawaetz, *Phys. Rev.* **4B**, 3460 (1971).
6. M. L. Cohen and T. K. Bergstresser, *Phys. Rev.* **141**, 789 (1966).
7. S. H. Groves, T. C. Harman, and C. R. Pidgeon, *Solid State Commun.* **7**, 451 (1971).

6

Pseudopotentials and Charge Densities

One of the most striking facts of nature is the repetition of chemical properties of the elements as the atomic number Z_A is increased. This repetition is the basis of the periodic table of the elements, and it is obviously desirable to make the quantum-mechanical basis for this repetition an explicit feature of any theoretical method for describing the electronic structure of solids. Although a number of different ways of doing this have been used successfully for atoms, one way has proved to be of especially great value for covalent crystals. This approach replaces the true potential of the atom by a pseudopotential,[†] which is defined in such a way as to provide both an accurate and a simple way for describing bonding in the crystal and relating this bonding to the original atomic properties. In this approach the atomic properties are regarded as given, either from experimental atomic spectra or from other theoretical studies of atomic properties, and emphasis is placed on how small changes in these atomic properties change the crystalline bond.

[†] For a more complete discussion of the mathematical properties of pseudopotentials and valence electron screening, see Phillips [1].

ATOMIC WAVE FUNCTIONS

The periodicity of chemical properties arises, of course, from the periodic building up of the shell structure of the atom, as indicated in Fig. 2.1. With the completion or closure of each row of the periodic table by a rare-gas atom (e.g., the Si row ends with the Ar closed-shell structure), a new row begins, and only the valence electrons outside the closed shell or rare gas core are chemically active. Thanks to this shell structure valence electron wave functions have an oscillatory character in the core region which is complementary to that of the core wave functions themselves. For example, a 2s atomic wave function has its node or zero near the point r_{1s} where the 1s atomic function reaches its maximum (antinode). Similarly, the 3s wave function has two nodes, one near the antinode of the 1s wave function, and one near the outer antinode of the 2s wave function. This behavior is a consequence of the shell structure of the atom and of the mathematical requirement that each nl wave function must be orthogonal to each $n'l$ wave function, $n' < n$. One may summarize the situation by saying that the oscillations of the valence wave function are out of phase with those of the inner core wave functions.

ATOMIC PSEUDOPOTENTIALS

According to the uncertainty principle, the wavelengths of these oscillations are related to the average momenta of the core electrons, and hence to their average kinetic energy. Within each shell of principal quantum number n', the potential seen by an electron in that shell is approximately given by $-Z_{eff}(r)e/r$, where Z_{eff} is given approximately by the total charge of the nucleus, Z_A, minus the charge of the electrons in shells n'', where $n'' < n'$. This means that the attractive potential experienced by the core electrons is much greater than that experienced by the valence electrons. According to the virial theorem, the kinetic energy of the core electrons will also be large, which means that

$$r_{n'l} \cong \hbar/p_{n'l} \qquad (6.1)$$

is small.

Because the valence electrons oscillate rapidly in each shell n' with a wave length of order $r_{n'l}$, in the core region they have high kinetic energies. One can remove the core oscillations or "wiggles" of the valence electron wave function ψ_{nl}, leaving only a smooth wave function φ_{nl}, the pseudo wave function. Similarly the atomic potential

$$V(r) = rZ(r)e \qquad (6.2)$$

is replaced by a pseudopotential

$$V_{\mathrm{p}}(r) = rZ'(r)e \qquad (6.3)$$

constructed so that it yields the pseudo wave functions φ. The pseudo wave function $\varphi_{nl}(r)$ is the same as the atomic wave function $\psi_{nl}(r)$ for r outside the core region, but as $r \to 0$ the pseudo wave function $\varphi_{nl}(r)$ tends to zero like r^l. We say that the radial nodes of $\psi_{nl}(r)$, which occurred near the antinodes of $\psi_{n'l}(r)$, have been removed from $\varphi_{nl}(r)$.

How does $V_{\mathrm{p}}(r)$ or $Z'(r)$ compare with $V(r)$ or $Z(r)$? In the core region as $r \to 0$, $Z(r)$ tends to Z_{A}, the atomic number of the element. On the other hand, a smooth wave function is one which sees almost zero potential, and this is what we expect for $Z'(r)$ in the core region. This behavior is shown for the 3s valence wave function φ_{3s} in Fig. 6.1, which shows $Z'(r)$ and $Z(r)$ versus r for Si^{3+}. Because of the particular mathematical transformation used, some oscillations in $Z'(r)$ in the core region are still evident near the antinodes of the 1s and 2s wave functions. For practical purposes, $V_{\mathrm{p}}(r)$ is usually treated as a constant as $r \to 0$. Ordinarily this constant is chosen as zero, because this makes φ even smoother in the core region.

CRYSTAL POTENTIAL

In the free atom each valence electron interacts with the Coulomb potential of the nucleus and the Coulomb potentials of the other electrons. Because of the shell structure of the atom it is convenient to divide the electrons into two groups, those in the atomic core, and the valence electrons in the last partially filled shell. The total number of core electrons is denoted by Z_{C}, the atomic number by Z_{A} and the number of valence electrons by Z. The net charge of the atomic core is $+Ze$, where

$$Z = Z_{\mathrm{A}} - Z_{\mathrm{C}}. \qquad (6.4)$$

In the crystal the electron charge density of the atomic cores is changed but little compared to the free atom. Thus we represent the crystal potential as a sum of ion core pseudopotentials, denoted by $V_{\mathrm{p}}{}^i$, and valence electron Coulomb potentials, denoted by V_{s} (see Fig. 6.2). The subscript s here is used to indicate that the valence electrons in the crystal are attracted to the ion cores, and in a first approximation act to shield or screen the ion core potentials.

An important practical point here is that this screening is relatively ineffective, so that on the average $V_{\mathrm{p}}{}^i$ is stronger than V_{s}. By studying atomic energy levels, for example, J. C. Slater concluded in 1930 that each s–p valence electron screens other s–p valence electrons in its own shell

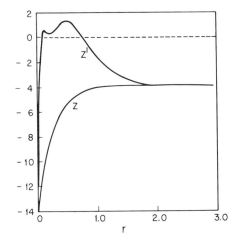

Fig. 6.1. The actual atomic potential of a Si^{4+} ion $V(r) = r\,Z(r)e$ compared with the pseudopotential $V_p(r) = r\,Z'(r)e$. For convenience it is $Z(r)$ and $Z'(r)$ that are actually shown plotted against r (in units of a_0). Outside the core both are equal to -4, corresponding to the Coulomb potential of the $+4$ ion. For $r/a_0 \lesssim 1$, $Z'(r)$ is practically zero, whereas $Z(r) \to 14 = Z_A(\text{Si})$ as $r \to 0$. Thus $Z'(r)$ defines the valence properties of the Si^{4+} ion.

from the nucleus by about 0.3e. That is, each valence electron is about 30% effective in reducing the core Coulomb potential seen by similar valence electrons. This is because the valence electron clouds are spread out, and a similar screening ineffectiveness is observed in crystals.

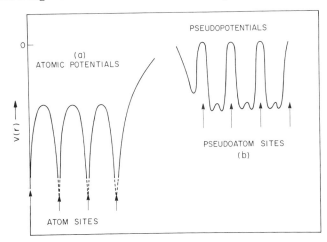

Fig. 6.2. (a) The crystal potential V as a superposition of atomic potentials; (b) the crystal pseudopotential V_p as a superposition of atomic pseudopotentials.

According to our discussion, the form of $V_p{}^i$ is given by

$$V_p{}^i = 0 \qquad \text{for} \quad 0 \leq r \leq r_c, \qquad (6.5)$$

$$V_p{}^i = -Ze^2/r \quad \text{for} \quad r_c \leq r. \qquad (6.6)$$

Our only task is to choose the value of r_c which is appropriate to a given atom. This can be done in several ways. One way is to use atomic energy levels, and adjust r_c so as to reproduce these levels. For our purposes here in studying crystals, we assume this has been done, so that $V_p{}^i$ is known. Because $V_p{}^i$ is much larger than V_s, one can make a guess at V_s and estimate it, for instance, from electronic screening potentials in atoms. One then uses $V_p{}^i + V_s$ obtained in this way to solve the Schrödinger equation for the valence-electron wave functions in the crystal. These will be slightly different from those of the free atoms, so a new value of V_s is calculated. Solving the Schrödinger equation again, we find that the new crystal wave functions differ only very slightly from those found in the first guess, so that one is usually justified in stopping at this point.

CRYSTAL WAVE FUNCTIONS

The crystal energy bands obtained in this way turn out to be close to those of free electrons, as we saw in Chapter 5. And, not surprisingly, the crystal wave functions are found to be made up predominantly of a few plane waves, because that is the form that the wave function takes in the free electron limit $(Z = 0)$. It is therefore desirable to study the interactions of plane waves with pseudoatoms, that is, the scattering of plane waves by atomic pseudopotentials. A plane wave of the form $\exp(i\mathbf{k}\cdot\mathbf{r})$ has a momentum $\mathbf{p} = \hbar\mathbf{k}$, and one therefore tends to describe such scattering events in terms of $\mathbf{k}$ or $\mathbf{p}$ space, rather than the more familiar language of $\mathbf{r}$ space. Ultimately, of course, the two descriptions are equivalent, but the $\mathbf{k}$ space one is actually simpler.

An electron wave in the crystal is scattered by pseudoatoms in a way analogous to the way X-rays are scattered by real atoms. The X-rays are scattered by lattice planes, from a state with initial wave number $\mathbf{k}$ to one with final wave number $\mathbf{k}'$. The difference between $\mathbf{k}'$ and $\mathbf{k}$ must be a reciprocal lattice vector $\mathbf{G}$,

$$\mathbf{k}' = \mathbf{k} + \mathbf{G}, \qquad (6.7)$$

according to Bragg's law. Similarly, an electron wave of momentum $\mathbf{p}$ is scattered to another wave of momentum $\mathbf{p}'$ with

$$\mathbf{p}' = \mathbf{p} + \hbar\mathbf{G}. \qquad (6.8)$$

The amplitude of scattering is determined by the matrix element $\langle \mathbf{p}' \mid V_p \mid \mathbf{p} \rangle$. Here $\mid \mathbf{p} \rangle$ represents the initial plane wave state and $\langle \mathbf{p}' \mid$ represents the final one, while V_p is the total crystal pseudopotential, summed over all the atoms in the crystal.

PSEUDOATOM FORM FACTORS

The expression for the scattering amplitude can be greatly simplified if one assumes that V_p can be written as a superposition of spherically-symmetric atomic pseudopotentials. This is true for $V_p{}^i$, the largest part of V_p, and it is approximately true for V_s, the valence electron potential, most of which is derived from screening charge concentrated about each atom. (Some valence charge is concentrated in bonds, but this is only a small part of V_s, which in turn is a small part of $V_p{}^i$. We therefore neglect it at present.) In this case one can write

$$\langle \mathbf{p}' \mid V_p \mid \mathbf{p} \rangle = \sum_\alpha V_p{}^\alpha(\mathbf{G}) S_\alpha(\mathbf{G}), \qquad (6.9)$$

where $S_\alpha(\mathbf{G})$ is the structure factor used to describe X-ray scattering from atoms of type α,

$$S_\alpha(\mathbf{G}) = \sum_{\mathbf{R}^\alpha_n \text{ in unit cell}} \exp(i\mathbf{G}\cdot\mathbf{R}_n{}^\alpha), \qquad (6.10a)$$

and $V_p{}^\alpha(\mathbf{G})$ is the pseudoatom form factor defined by

$$V_p{}^\alpha(\mathbf{G}) = \Omega^{-1} \int \exp[i(\mathbf{k}' - \mathbf{k})\cdot\mathbf{r}] V_p{}^\alpha(\mathbf{r}) \, d^3\mathbf{r}, \qquad (6.10b)$$

where $V_p{}^\alpha(\mathbf{r})$ is the atomic pseudopotential, described, for example, by Eqs. (6.5) and (6.6). In Eq. (6.10b) the integral is taken over all space, and Ω is the atomic volume.

From the way we have defined $V_p(\mathbf{G})$, through Eq. 6.10b, it appears that $V_p(\mathbf{G})$ is unique. A precise definition requires certain mathematical refinements which need not concern us here, but which the interested reader can pursue elsewhere [1]. The most important of these arise because V_p may depend not only on $\mathbf{p} - \mathbf{p}'$ but also on $\mathbf{p} + \mathbf{p}'$. This additional dependence is suppressed by concentrating attention on interactions where $\mathbf{p}$ and $\mathbf{p}'$ have magnitudes near $p_F = \hbar k_F$, the Fermi momentum. We recall from Chapter 2 that k_F is defined by the average number of valence electrons Z per atomic volume Ω, so that Ω enters Eq. 6.10b both explicitly and implicitly. The condition that $\mid \mathbf{p} \mid$ and $\mid \mathbf{p}' \mid$ be near p_F is equivalent to saying that in discussing the energy gap E_g between bonding and antibonding states we are primarily concerned with momenta in the range

$$p_F(1 - E_g/4E_F) \lesssim p \lesssim p_F(1 + E_g/4E_F). \qquad (6.11)$$

This is the range that is most important, according to the discussion in Chapter 5.

There are several ways that pseudoatom form factors can be determined in practice. A good first approximation can be obtained from studying free atoms. So much is known about semiconductors experimentally that one often sees the pseudoatom form factors determined from experiment. This is sometimes called the empirical pseudopotential method (EPM). The value of EPM is especially great because it turns out that, like elemental electronegativities, pseudoatom form factors are transferrable. That is, a pseudoatom form factor $V_p(A)$ can be determined from the properties of crystal A, another form factor $V_p(B)$ fixed by the properties of crystal B, and then the properties of compound AB can be predicted from $V_p(A)$ and $V_p(B)$. By this method pseudoatom form factors [2] have been determined for most of the elements forming semiconductor compounds, as listed in Fig. 2.1.

The only points at which the form factor $V_p(\mathbf{q})$ need be known in the crystal are those for which $\mathbf{q} = \mathbf{G}$, where $\mathbf{G}$ is one of the first two or three reciprocal lattice vectors. These two or three numbers can be determined very accurately from experiment and then compared with the free atom form factor $V_p(q)$. When this was done for Ge, it was found that the EPM values for $V_p(\mathbf{G})$ in the crystal could be defined as a smooth curve, similar to the free atom curve, but passing through the crystalline values $V_p(\mathbf{G})$, as shown in Fig. 6.3. The dashed curve can be used now in other crystals, where different values of $\mathbf{G}$ may be needed. Whatever corrections may be responsible for the differences between the two curves of Fig. 6.3, $V_p(\mathbf{G})$ apparently changes little on passing from one crystal structure to another. This should not be too surprising, because the greatest change on passing from the gaseous to the crystalline state is one of density, and all covalent crystals exhibit similar densities on this scale.

Many theories of the electronic structures of molecules or crystals are atomistic in character. This is obviously the case in molecular orbital calculations which represent the molecular wave function as a linear combination of atomic orbitals. Also many semiempirical theories of molecular structure utilize atomic ionization energies and electron affinities to estimate relative electronegativities of different atoms. When one considers the crystalline state, however, these atomistic concepts represent a poor point of departure, because the atoms are compressed and the valence electrons are delocalized. (Any given electron is equally likely to be in any unit cell of the crystal.) It is important to realize that *both* of the pseudoatom form factors

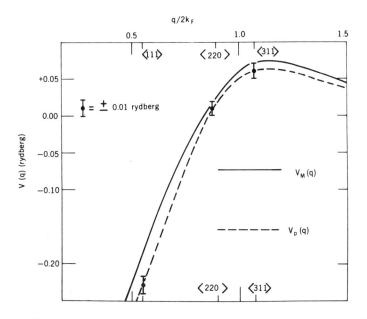

Fig. 6.3. Comparison of pseudoatom form factors for Ge. The solid curve is based on free-atom energy levels; the dashed curve passes through the indicated crystalline values for the reciprocal lattice vectors $\mathbf{G} \sim (1,1,1)$, $(2,2,0)$ and $(3,1,1)$.

shown in Fig. 6.3 are truly crystalline and not atomistic in character. What these form factors describe are *atoms in crystals.*

Because the EPM curve was determined empirically from crystalline energy values, it obviously can be described in terms of "atoms in crystals," but this may not be so clear for the "free atom" curve, which is based on free-ion term values. After the free-ion form factor has been determined, the "free-atom" form factor is determined by immersing the ion in a free-electron gas with density equal to that of the valence electrons in the corresponding elemental crystal. The resulting screening charge is then computed to first order in $V_p{}^i$. The resulting pseudoatom form factor represents the potential obtained from the ion core potential plus just enough screening charge to neutralize the atom. Thus the "free-atom" form factor corresponds to an *atom in a nearly free electron gas.* This is not quite the same as the atom in the crystal, and this accounts for most of the difference between the two curves shown in Fig. 6.3.

The language of pseudoatom form factors is convenient for discussing changes in crystalline binding. We now compare metallic and covalent binding in these terms, in order to appreciate the factors responsible for the great strength of the covalent bond.

METALLIC BINDING

In nontransition metals, the free-electron approximation has been found to give quite an accurate account of the electronic structure and energy bands $E_n(\mathbf{k})$. The reason for this in the case of a close-packed metal such as Al can be explained quite easily in terms of the pseudopotential form factor. As shown in Fig. 6.4a, the first two reciprocal lattice vectors (labeled $\mathbf{G}_1$ and $\mathbf{G}_2$) fall close to the point q_0 at which $V_p(q) = 0$. Therefore $V_p(\mathbf{G}_1)$ and $V_p(\mathbf{G}_2)$ are both small compared to a typical kinetic energy such as E_F. Thus V_p can be treated to first order, and it will affect the energy bands

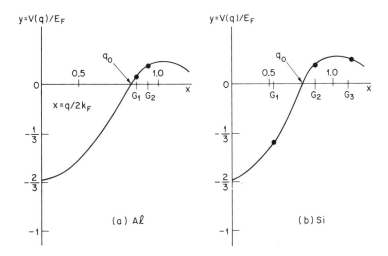

Fig. 6.4. The geometry of reciprocal lattice vectors and pseudoatom form factors. In (a) we have the situation which generally obtains in close-packed metals. Here the first few reciprocal lattice vectors fall near the node in the pseudoatom form factor caused by cancellation between the attractive screened Coulomb potential of the ion core, and the effective repulsive potential produced by the exclusion principle. If a reciprocal lattice vector $\mathbf{G}$ falls in the region where the pseudopotential form factor is repulsive [as is the case here for $V(\mathbf{G}_1)$ and $V(\mathbf{G}_2)$], then the ordering of s and p levels normally found in atoms is reversed in the crystal. For example, the states

$$\exp(i\mathbf{G}_1 \cdot \mathbf{r}/2) \pm \exp(-i\mathbf{G}_1 \cdot \mathbf{r}/2)$$

correspond respectively to s-like and p-like states, with the p-like state oriented parallel to $\mathbf{G}_1$. The two states differ in energy by $2V(\mathbf{G}_1)$, the s-like one being higher if $V(\mathbf{G}_1) > 0$. This is the situation that occurs in metals like Al.

In (b), on the other hand, because of the larger unit cell, there is one small reciprocal lattice vector $\mathbf{G}_1$ such that $V(\mathbf{G}_1)$ is large in magnitude and negative in sign. It is this Fourier component of the pseudopotential that gives rise to covalent bonding.

only when two free-electron states (such as $\mathbf{k} = \mathbf{G}_1/2$ and $\mathbf{k}' = -\mathbf{G}_1/2$) have very nearly the same kinetic energy and differ by a reciprocal lattice vector (such as $\mathbf{G}_1$). The fraction of such states itself is of order $V_p(\mathbf{G})/E_F$, and is therefore small.

COVALENT BINDING

In the diamond-type crystals, the position of the Bragg scattering vectors $\mathbf{G}_i$ differs dramatically from that of the close-packed metals. The geometrical reason for this is that the prototypical close-packed metallic structures are monatomic, while the diamond-type structures are diatomic. As a result, there is one set of reciprocal lattice vectors, the ones normal to (111) lattice planes, which are much smaller in magnitude than any of the reciprocal lattice vectors in the close-packed structure. (This is because the diatomic unit cell is larger, and hence by the uncertainty principle, the unit vectors in $\mathbf{k}$ space are smaller.) The position of these vectors, labeled $\mathbf{G}_1$, on the pseudoatom form factor curve for Si is shown in Fig. 6.4b. Also shown are the positions of $\mathbf{G}_2$ and $\mathbf{G}_3$, the next two reciprocal lattice vectors.

From Fig. 6.4b, we see that $V_p(\mathbf{G}_1)$ is about 0.4 E_F, while $V_p(\mathbf{G}_2)$ and $V_p(\mathbf{G}_3)$ are of order 0.1 E_F. Thus $\mathbf{G}_2$ and $\mathbf{G}_3$ fall close to the nodal point q_0 of $V_p(q)$ for Si, just as $\mathbf{G}_1$ and $\mathbf{G}_2$ did for Al. This means that $V_p(\mathbf{G}_2)$ and $V_p(\mathbf{G}_3)$ can be treated to first order, while $V_p(\mathbf{G}_1)$ must be handled more carefully.

A model of the covalent energy gap E_h in diamond-type solids in terms of the effect of V_p on energy states near $E = E_F$ can be developed [2, 3]. The formula which describes E_h in terms of V_p is

$$\tfrac{1}{2}E_h = V_p(220) + [V_p(111)]^2/\Delta T, \tag{6.12}$$

$$\Delta T = \hbar^2(2\pi/a)^2/2m, \tag{6.13}$$

where a is the cubic lattice constant. The reciprocal lattice vectors $\mathbf{G}_1$ and $\mathbf{G}_2$ in Fig. 6.4b are the same as those abbreviated by (111) and (220), respectively, in Eq. (6.12).

It is interesting to evaluate the first and second terms on the right-hand side of Eq. (6.12) explicitly for Si. The values of $V_p(\mathbf{G})$ can be obtained either from atomic calculations or from empirical fits to energy bands (EPM), as shown in Fig. 6.3. In either case one obtains $V_p(220) \cong 0.6$ eV, $V_p(111) \cong -2.7$ eV, and from the cubic lattice constant, ΔT is found to be 5.1 eV. Thus the second term is 1.4 eV, or more than twice the first term, which is only 0.6 eV. The sum gives $E_h \cong 4.0$ eV, in fairly good agreement with the value of 4.8 eV given in Table 2.2 considering the oversimplified and approximate nature of Eq. 6.12. Most importantly, the term of metallic character, $V_p(220)$, is small compared to the nonlinear covalent term $[V_p(111)]^2/\Delta T$.

IONIC BINDING

The expression for E_h given in Eq. (6.12) can be generalized to describe the "complex" energy gap,

$$E_g = E_h + iC. \tag{6.14}$$

Of course, in practice we are always concerned only with the magnitude $|E_g|$, which is a real energy. But as we saw in Eq. (6.9), the matrix elements of the crystal potential depend on the pseudoatom form factor (which is real) and on the structure factor (which may be complex). In an elemental diatomic crystal the structure factor takes the form

$$S(\mathbf{G}) = \exp (i\mathbf{G} \cdot \mathbf{R}_A) + \exp (i\mathbf{G} \cdot \mathbf{R}_B), \tag{6.15}$$

where $\mathbf{R}_A$ and $\mathbf{R}_B$ mark the positions of atoms A and B. By choosing our origin of coordinates halfway between the two atoms in the unit cell, we can make $\mathbf{R}_B = -\mathbf{R}_A$. Then $S(\mathbf{G})$ is real and $C = 0$.

What happens when the two atoms in the unit cell are different? Then (6.9) takes the form,

$$\langle \mathbf{p}' \mid V_p \mid \mathbf{p} \rangle = V_p^A(\mathbf{G}) \exp (i\mathbf{G} \cdot \mathbf{R}_A) + V_p^B(\mathbf{G}) \exp (i\mathbf{G} \cdot \mathbf{R}_B) \tag{6.16}$$

where V_p^A and V_p^B denote the pseudoatom form factors of atoms A and B, respectively. With the definitions

$$V_s(\mathbf{G}) = V_p^A(\mathbf{G}) + V_p^B(\mathbf{G}), \tag{6.17}$$

$$V_a(\mathbf{G}) = V_p^A(\mathbf{G}) - V_p^B(\mathbf{G}), \tag{6.18}$$

and the choice of origin so that

$$\mathbf{R}_B = -\mathbf{R}_A \tag{6.19}$$

as before, Eq. (6.16) becomes

$$\langle \mathbf{p}' \mid V_p \mid \mathbf{p} \rangle = 2V_s(\mathbf{G}) \cos \mathbf{G} \cdot \mathbf{R}_A + 2iV_a(\mathbf{G}) \sin \mathbf{G} \cdot \mathbf{R}_A. \tag{6.20}$$

For $\mathbf{G} \sim (1, 1, 1)$, it turns out that the cos and sin factors in (6.20) are equal, while for $\mathbf{G} \sim (2, 2, 0)$, the sin factor is zero. Thus the appropriate generalization of Eq. (6.12) to the partially ionic case where $V_A \neq V_B$ is [3]

$$\tfrac{1}{2}E_h = V_s(220) + [V_s(111)]^2/\Delta T, \qquad (6.21)$$

$$\tfrac{1}{2}C = 2V_s(111)V_a(111)/\Delta T. \qquad (6.22)$$

The form of Eq. (6.21) has been chosen to reflect the fact that in an isoelectronic series like Ge, GaAs, ZnSe, the bond length d changes negligibly, as does V_s. Because $E_h = E_h(d)$, E_h must depend only on V_s as indicated by Eq. (6.21).

The description of ionic effects in terms of pseudopotentials can be checked in the same way that the validity of (6.12) was tested. For GaAs the pseudopotential value of C is 2.5 eV, compared to the spectroscopic value of 2.9 eV. For ZnSe the respective values are 5.0 eV and 5.7 eV. Just as we found with E_h, the pseudopotential values agree rather well with the spectroscopic ones, although they are in the cases mentioned always 10–20% smaller. This presumably reflects algebraic simplifications used to derive (6.12), (6.21), and (6.22) [3] .

There is an aspect of equation (6.22) which deserves some discussion. Because V_a doubles on going from GaAs ($\Delta Z = 5 - 3 = 2$) to ZnSe ($\Delta Z = 6 - 2 = 4$) while V_s remains unchanged, it is necessary that E_h (which is unchanged) depend only on V_s, while C (which also doubles on going from GaAs to ZnSe) be proportional to V_a. Together with the behavior of $S(\mathbf{G})$, this virtually determines the forms of (6.21) and (6.22), apart from factors of order unity.

The reader should not be greatly surprised that a correspondence can be established between the bond parameters E_h and C and the pseudopotential band parameters $V_s(\mathbf{G})$ and $V_a(\mathbf{G})$. The latter, as we shall see in the next chapter, are determined by fitting various peaks and edges in $\epsilon_2(\omega)$, while the former are derived from $\epsilon_1(0)$. Because of the Kramers–Kronig relation, equation (7.3), the values of E_h and C represent averages of the effects of $V_s(\mathbf{G})$ and $V_a(\mathbf{G})$. Approximately the same averages are involved in all tetrahedrally coordinated semiconductors, and thus one would expect to find correspondences such as are represented by Eqs. (6.21) and (6.22).

The numerical examples of the correspondence between pseudopotentials derived from band theory and average energy gaps derived from dielectric bond theory have been straightforward so far. This is because we have confined ourselves to isoelectronic sequences such as Ge, GaAs, and ZnSe where all atoms come from the same row of the periodic table (see Fig. 2.1).

In isoelectronic sequences bond lengths are constant, so that the values of E_h and V_s are virtually the same for A^2B^6 as for A^3B^5, while the values of C and V_a are twice as great for A^2B^6 as for A^3B^5, in accordance with the expectations of chemical valence theory.

A more demanding test of the accuracy of each theory is posed by comparing two different compounds with $N = 3$ composed of atoms from two different rows. For example, referring to Fig. 2.1, we see that Ga and As belong to the Ge row, while In and Sb belong to the Sn row. The two compounds GaSb and InAs thus are of the form A^3B^5, with one atom from the Ge row and one atom from the Sn row each. However, according to Table 2.2 GaSb is only 26% ionic, while InAs is 36% ionic. The reason for this is the factor $(Z_A/r_A - Z_B/r_B)$ which occurs in the formula (2.24) for C. With $Z_A = 3$ for the cation and $Z_B = 5$ for the anion, C^2 will be larger for the compound InAs, where the anion is smaller than the cation, i.e., $r_B < r_A$, than it is for GaSb, where $r_A < r_B$. Indeed, according to Table 2.2, the values of E_h in these two compounds are the same within 3%, but $C(\text{InAs})$ = 2.7 eV compared to $C(\text{GaSb})$ = 2.1 eV, a difference of about 25%. The values [2] of $V_a(111)$ in InAs and GaSb adjusted to fit optically determined energy band differences are 0.08 Ry, and 0.06 Ry, respectively, which gives a ratio in good agreement with the ratio of the bond values of C in the same two compounds. The absolute agreement between E_h and C, on the one hand, and $V_s(111)$ and $V_a(111)$ on the other, is comparable to that quoted earlier for GaAs. The absolute agreement, however, is probably less significant than the ability of both theories to describe the chemical trends from GaSb ($f_i = 0.26$) to InAs ($f_i = 0.36$).

SEMICONDUCTOR WAVE FUNCTIONS

In order to obtain accurate pseudowave functions for the entire valence band one proceeds as follows. The pseudowave function φ, which is the smooth or (radial) nodeless part of the true wave function ψ, is expanded in a three-dimensional Fourier series, which has the form

$$\varphi_k(r) \sim \sum_G a_G \exp [i(\mathbf{k} + \mathbf{G}) \cdot \mathbf{r}]. \qquad (6.23)$$

The Schrödinger equation can be written as

$$\sum_G \langle \mathbf{k} + \mathbf{G}' \mid H \mid \mathbf{k} + \mathbf{G} \rangle a_G = E a_{G'}, \qquad (6.24)$$

where the Hamiltonian H consists of the kinetic energy and the sum of the pseudoatom potentials. If n reciprocal lattice vectors $\mathbf{G}$ are used in the ex-

pansion (6.23), then (6.24) is a set of n simultaneous equations for the eigenvalues $E_n(\mathbf{k})$ and the eigenfunction Fourier coefficients $a_{\mathbf{G}}{}^n$.
In metals such as Al, where V_p is weak for the reasons shown in Fig. 6.4a, near $E = E_\mathrm{F}$ no more than half a dozen terms are used in the expansion (6.23), and (6.24) can be solved by hand. In crystals where bonding is important, and $V_\mathrm{p}(\mathbf{G})$ is large for at least one $\mathbf{G}$, $n = 100$ or more, and in that case one must certainly use a computer. Moreover the coefficients $a_{\mathbf{G}}{}^n$ vary with $\mathbf{k}$ in the Brillouin zone, as do the energy bands $E_n(\mathbf{k})$.

We have already looked at the energy bands $E_n(\mathbf{k})$ for several semiconductors in the preceding chapter. Now that we understand the nature of pseudoatom potentials, it is appropriate to look at valence charge densities in semiconductor crystals.

PSEUDOCHARGE DENSITIES

Because the forces that determine crystal structure are the Coulomb forces between positive nuclei and negative electrons, one can obtain insight into many structural properties by studying electronic charge densities. However, before constructing the total charge density in the crystal, we should pause to consider the implications of quantum mechanics for this problem.

There are two quantum-mechanical effects that prevent us from directly applying classical intuition to this problem. The first has to do with the fact that it is wave function amplitudes that satisfy the wave equation, not intensities. When ρ, the charge density of electrons, varies rapidly in space, one can attempt to develop statistical theories (of the Thomas–Fermi type) to describe such variations. According to such theories, one finds kinetic energy terms proportional to $(\nabla\rho)^2$ associated with such variations. Unfortunately, the conditions for validity of the statistical theories always reduce to the statement that the charge densities vary slowly in an atomic wavelength, whereas it is just variations over distances less than or equal to atomic wave lengths that are of interest to us. And over these distances we find that interference terms, which are omitted from statistical or semiclassical theories, can be quite important. In fact, interference terms turn out to make essential contributions to covalent bonds in semiconductors.

The second quantum effect which is very important is the exclusion principle. Without it all the valence electrons would collapse into 1s states about the atomic nuclei. If electrons did not even have to satisfy the wave equation, they would all spiral into the nearest nucleus, and matter would collapse. This is called the ultraviolet catastrophe, because according to classical mechanics, as the electrons spiraled into nuclei, they would have emitted ultraviolet radiation. Similarly, without the exclusion principle, we would have a hydrogenic catastrophe, with all atoms resembling a hydrogen atom, only with different values of nuclear charge.

The physical implications of the exclusion principle are not ended after the inner cores of the atoms have been built up. When the atom cores are displaced, as in thermal vibrations, or rearranged, as in a phase transition, the effects of the exclusion principle follow the nuclear motion, just as the core electrons do. If the atoms were well separated, the effects of the exclusion principle could be treated by conventional multiple scattering theory. This theory uses phase shifts and is designed to describe a total charge density which is a superposition of spherically symmetric charge densities centered on the atoms. (The charge densities on each atom may, nevertheless, have a different radial dependence than those of the corresponding free atoms.) However, as we shall see, in semiconductors an essential part of the charge density is centered in bonds, and not in atoms. One of the great advantages of pseudopotential theory is that it supplies an accurate, simple description of the accumulation of electronic charge in these bonds.

Why does pseudopotential theory work so well in describing covalent bonds, when it is still based on pseudoatom form factors? The answer is that the form factors can be adjusted to reproduce all the energy levels of the highest valence and lowest conduction bands *of the crystal*. These bands, which are those that lie nearest the Fermi energy, are exactly the ones that are associated with low-energy optical transitions. The low-energy optical transitions, discussed in detail in Chapter 7, are the ones that determine the polarizability. The charge redistribution that occurs in low-energy optical transitions is of the same kind that occurs in thermal vibrations and even, to a considerable extent, in phase transitions (such as gray Sn–white Sn) which are heralded by lattice instabilities (Chapter 4). By contrast, phase shifts describe core states, and thereby relate to core–conduction band transitions. These make a negligible contribution to the polarizability and to most other quantities of structural interest. There is, however, something to be learned from the core states about charge transfer.

ATOMIC CHARGES

Valence electrons penetrate the ion core only slightly. However, when they do so (which is usually less than 10% of their probability distribution), their charge repels the charge of the core electrons. Alternatively stated, the valence charge in the core region acts to screen partially core electron–nuclear attractions, thereby reducing core–binding energies. If we now remove some valence charge from the atom, the core electrons will be attracted more strongly to the nucleus, and their binding energies will become more negative (larger in magnitude). These energies can be measured to within a few tenths of an eV (or better) relative to vacuum by photoelectron spectroscopy. This process has been given the acronym ESCA (electron spectroscopy for chemical analysis).

At this writing the quantity of ESCA data which is available for solids, and in particular for semiconductors, is still quite limited. However, on the basis of the results obtained to date, it appears that shifts in energies of least-bound core electrons provide the least ambiguous measure of atomic charge states. The reason for this is that most other measures of charge states involve geometrical assumptions, such as dividing up the crystal volume into disjoint atomic volumes. Such divisions are necessarily arbitrary, and may not reflect the charge-transfer contribution to heats of formation (Chapter 8).

Energy shifts of the least-bound core levels of A^2B^6 semiconductors (A = Zn, Cd, Hg; B = O, S, Se, Te) are shown in Fig. 6.5. The sign of the energy shift corresponds to charge transfer from the A atom to the B atom. The energy shifts are measured relative to the core levels of the correspond-

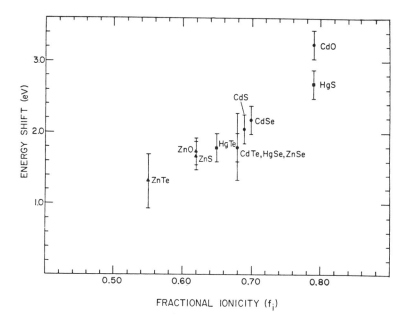

Fig. 6.5. Energy shifts of least-bound core levels plotted as a function of the spectroscopic ionicity f_i defined by Eq. (2.28) [from C. J. Vesely and D. W. Langer, *Phys. Rev.* **4B**, 451 (1971)].

ing A metal, so that, roughly speaking, the A atoms have fewer valence electrons in the core region in the semiconductor crystal than they would have had in the metal. Because the metals are nearly close-packed (almost 12-fold coordinated) compared to fourfold coordination in the A^2B^6 crystal, some of the shift arises from changes in valence-electron density and nearest-neighbor bond lengths. However, superimposed on this is a charge transfer which arises from the ionic character of the A^2B^6 bond. This transfer appears to be approximately linear in the spectroscopic ionicity defined by Eq. (2.28).

BOND CHARGES

The total pseudocharge density in the crystal can be obtained by summing over a sample of states in the valence band labeled by $\mathbf{k}$ and a band index n which runs over the four valence bands. This defines a valence pseudocharge density ρ:

$$\rho = \sum_{\mathbf{k},n} |\varphi_{\mathbf{k},n}(\mathbf{r})|^2, \tag{6.25}$$

where $\varphi_{\mathbf{k}}$ has been expanded in plane waves, as in Eq. (6.23). The crystal electronic charge density is then given by the Fourier series

$$\rho(\mathbf{r}) = \sum_{\mathbf{G}} \rho_{\mathbf{G}} \exp(i\mathbf{G}\cdot\mathbf{r}), \tag{6.26}$$

where $\rho_{\mathbf{G}}$ is given simply by

$$\rho_{\mathbf{G}} = 2Re \sum_{\mathbf{k},n} a^*(\mathbf{k}, n)a(\mathbf{k} + \mathbf{G}, n), \tag{6.27}$$

and the normalization of the number density is given by

$$\rho_0 = \sum_{\mathbf{k},n} |a_0(\mathbf{k}, n)|^2 = 8/\Omega, \tag{6.28}$$

where Ω is the volume of the unit cell in the crystal.

When the three-dimensional Fourier series (6.26) has been calculated, one is still left with the problem of displaying the results on the two-dimensional page. Two ways of doing this are helpful. Along the (111) bond directions one can label distance with x and plot $\rho(x)$. This is done in Fig. 6.6 for three isoelectronic compounds: $Ge(f_i = 0)$, $GaAs(f_i = 0.31)$, and $ZnSe(f_i = 0.63)$. The (111) direction is an interesting one, not only for the behavior of ρ in the bonding region between the two atoms, separated by a distance d, but also in the interstitial region where the separation of atoms is $3d$.

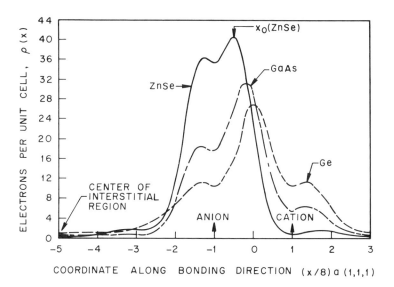

COORDINATE ALONG BONDING DIRECTION $(x/8)a\,(1,1,1)$

Fig. 6.6. Profiles of pseudocharge densities $\rho(x)$ for $\mathbf{r}$ along a (111) axis passing through the two atoms in the unit cell. Results are shown for Ge, GaAs, and ZnSe.

Study of the charge profiles shown in Fig. 6.6 reveals a number of interesting qualitative features. These are

1. With regard to the Fourier series (6.26), if we were to keep only one term, $\rho = \rho_0$, we would have a constant charge density, even in the interstitial regions. If we keep a few terms, the charge density in the interstitial region will be greatly reduced on the average, but there will still be oscillations in $\rho(x)$ leading to negative values of ρ, which are unphysical. Therefore the accuracy of the calculation, based on a limited number of terms in the series (6.26), can be determined by how closely $\rho(x)$ approaches zero in the interstitial region without actually becoming negative. Using this criterion we can see that the Fourier synthesis is very accurate.

2. There is a dramatic pileup of charge in the region between the two atoms. The volume of this region is small compared to the volume of the unit cell. The localization of charge is highly effective, however, and it gives rise to a sharp peak in charge density in the bonding region. This bond charge is the one that has been used in electrostatic pseudopotential models of the lattice vibrations of diamond-type lattices (Chapter 4).

3. In partially ionic crystals $(f_i \neq 0)$, the bond charge is displaced towards the B anion. The positions x_0 of the peaks in the bond profiles for Ge, GaAs, and ZnSe are plotted as functions of f_i in Fig. 6.7. An unsatisfactory feature of the charge profile displays is that one obtains no idea of

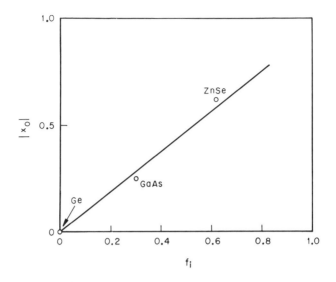

Fig. 6.7. The center x_0 of the bond charge densities shown in Fig. 6.6 plotted as a function of the spectroscopic ionicity f_i. The center is displaced towards the anion with increasing f_i.

the transverse dimensions of the bond charge. A (110) plane in the unit cell, indicated schematically in the inset to Fig. 6.8, passes through the two atoms in the unit cell and also through two atoms in adjacent cells, thereby displaying one full bond in the unit cell and halves of two other bonds. A topographical contour map can then be drawn to indicate levels of constant charge density, as shown in Fig. 6.8 for Ge. This map reveals several interesting features of the valence electron charge density.

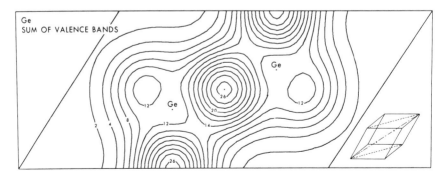

Fig. 6.8. A topographical map of the valence pseudocharge density of Ge in the (110) plane. The position of this plane in the unit cell is shown in the inset [from J. P. Walter and M. L. Cohen, *Phys. Rev.* **4B**, 1877 (1971)].

4. The bond charges dominate the topographical map, although they fill only a small portion of the volume of the unit cell. The symmetry of the bond axis implies axial symmetry of the bond charge (neglecting the effects of overlap with adjacent bonds), but the actual ellipsoidal distribution is nearly spherical, with a small elongation along the bond axis (slightly prolate ellipsoid).

5. One can analyze the contributions of each valence band separately to the total valence-electron charge density. The contribution to the bond charge increases as we go from the lowest energy band of s-like symmetry to the higher valence bands of p-like symmetry. This is what one would have expected, since the higher bands are closer to the energy gap between bonding and antibonding levels. The bond charge is quite prolate for the lowest band, but is oblate for the highest one. This corresponds to a transition, in the chemical language of atomic orbitals, from σ bonds in the lowest band to π bonds in the highest band, a result which is in good accord with conventional chemical usage in planar or linear hydrocarbon molecules.

PARTIALLY IONIC CHARGE DISTRIBUTIONS

In discussing Fig. 6.6 we have already seen that the center of the bond charge is displaced towards the anion in partially ionic semiconductors, and that the magnitude of this displacement increases with increasing ionicity. From topographical maps for crystals like GaAs (see Fig. 6.9) and ZnSe, one can see that the volume and magnitude of the bond charge decrease as more and more valence charge is localized around the anion. The volume and magnitude to be ascribed to the bond charge again are to

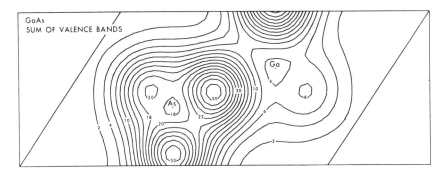

Fig. 6.9. A topographical map of the valence charge density of GaAs, given in units of c/Ω [from J. P. Walter and M. L. Cohen, *Phys. Rev.* **4B**, 1877 (1971)].

some extent arbitrary, and the results obtained depend to some extent on the definition used. However, if a common procedure is employed to define the background and to subtract it to obtain the bond charge Z_b, then the trend of Z_b as a function of f_i is unlikely to be greatly affected by the definition employed. In the results below ρ_0 is the value of ρ at the lowest level contour completely enclosing the bond charge peak. Then the bond charge Z_b is taken from an integral over the volume Ω_0 contained within this surface. Thus Z_b is given by

$$Z_b = \int_{\Omega_0} (\rho - \rho_0) \, d^3r. \tag{6.29}$$

The bond charge Z_b given by Eq. (6.29) is shown plotted against f_i in Fig. 6.10. The magnitude of the bond charge apparently goes to zero near $f_i = 0.8$.

By studying Fig. 6.7 and Fig. 6.10 we see that there are two possible electrostatic explanations for the fact that $A^N B^{8-N}$ compounds are stable with fourfold coordination for $f_i \leq F_i = 0.785$, but have sixfold coordination for $f_i \geq F_i$. One explanation is that the bond charge must lie between two nearest neighbors, since for $f_i > F_i$ the bond charge, according to Fig. 6.7, might well be centered on the interstitial side of the anion. The other explanation is that for the tetrahedral structure to be stable, Z_b must exceed a minimum value, which however is very close to zero (of order 0.02

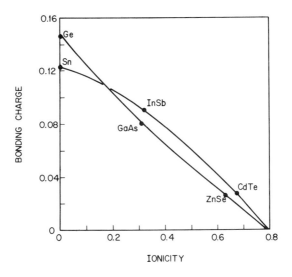

Fig. 6.10. The bond charge Z_b as a function of the spectroscopic ionicity f_i [from J. P. Walter and M. L. Cohen, *Phys. Rev.* **4B**, 1877 (1971)].

or less, in units of e). Because of the intrinsic ambiguities associated with the use of φ, the pseudowave function, to describe only part of the valence charge, and especially in the region of the anion cores for $f_i \gtrsim 0.7$, it may not be possible to decide between these two pictures. It does seem safe, however, to say that the stability of tetrahedral partially-ionic crystals is determined primarily by electrostatic forces connected with bond charges.

It is interesting to compare the bond charge trends with the softening of the tetrahedral structure with respect to shear. In Fig. 3.2 we saw that the ratio β/α of noncentral (bond-bending) forces to central (bond-stretching) forces decreased linearly with f_i, much as Z_b does. Therefore, as the bonds of $A^N B^{8-N}$ semiconductors became more ionic, they lose their directional stability and the bond charges become smaller, until finally at the critical ionicity F_i the low-density tetrahedral structure becomes unstable, and the phase transition to the higher-density NaCl structure takes place.

There are many similarities and differences between the valence-charge densities depicted in Figs. 6.6 and 6.8 and what one would have expected on the basis of hybridized atomic orbitals $\Psi_{sp^3}^A$ and $\Psi_{sp^3}^B$ of the kind discussed in Chapters 1 and 2. On the one hand the bond charges are much better localized than one would have expected simply from the overlap of atomic orbitals. The reason for this is that when the charge densities are constructed from superposition of many plane wave amplitudes, the resulting Fourier series is better able to represent shortwave length variations of $\rho(\mathbf{r})$. One can analyze the contributions to each Fourier amplitude $a_G(\mathbf{k}, n)$ in Eq. (6.23) in terms of powers of $V_A(\mathbf{G}_1)$, $V_A(\mathbf{G}_2)$, $V_B(\mathbf{G}_1)$, etc., and show that the localization of the bond charge arises largely because of contributions which are of second and higher order in the pseudoatom V's. The terms of first order in the pseudoatom V's on the other hand, correspond more nearly to the superposition of atomic charge densities which one would obtain from the hybridized atomic orbital approach.

One feature of the valence-electron charge density upon which the two methods agree is the charge accumulation about the anion in the partially ionic crystal. From Fig. 6.9 one is inclined to say, for example, that the ratio of charge about the As anion to that about the Ga cation is greater than the ratio of $5:3$ required for charge neutrality. This conclusion is in agreement with the results of the ESCA studies mentioned on p. 141.

CONDUCTION BAND STATES

While the valence band energies E^v_k always exhibit the behavior to be expected from bonding combinations of hybridized atomic sp³ orbitals, as we saw in the last chapter, the behavior of conduction band states is much

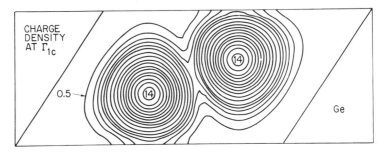

Fig. 6.11. Contours of constant pseudocharge density for the conduction band state Γ_{1c} (or $\Gamma_{2'c}$ if account is taken of inversion symmetry) in Ge. The contours correspond in the atomic orbital model to antibonding s states centered on the two atoms in the unit cell [figure courtesy of J. P. Walter and M. L. Cohen (unpublished)].

more complicated. Changes in chemical composition, pressure, or temperature produce quite different shifts in conduction band edges at different points $\mathbf{k}$ in crystal momentum space. The reasons for such large variations in behavior became apparent when one examines the spatial distributions of charge densities $\rho_{\mathbf{k}}^{c}$ of conduction band states $\psi_{\mathbf{k}}^{c}$.

For illustrative purposes we describe $\rho_{\mathbf{k}}^{c}$ for the states Γ_{1c}, L_{1c}, and X_{1c}, corresponding to $\mathbf{k} = 0$, and the centers of the (111) and (200) faces of the Brillouin zone, respectively. The topographical maps for Γ_{1c} and X_{1c} are shown in Figs. 6.11 and 6.12, respectively. These are the extreme cases, while the contours for L_{1c} lie about equally between those of Γ_{1c} and X_{1c}. The same is true of L_{1c} relative to Γ_{1c} and X_{1c} as a function of chemical composition, temperature, and pressure.

The symmetries of Γ_{1c}, L_{1c} and X_{1c} correspond, in atomic orbital language, to antibonding combinations of s atomic orbitals on the atoms A and B in

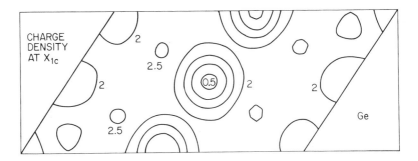

Fig. 6.12. Contours of constant pseudocharge density for the conduction band state X_{1c} in Ge. The contours show an almost uniform distribution of charge in the unit cell, which is quite contrary to expectations based on atomic orbital models.

the unit cell, represented formally by $a\Psi_A - b\Psi_B$. At $k = 0$ in the case of Ge, shown in Fig. 6.11, this means that $\rho = 0$ at the bonding site halfway between the two atoms, and this is seen to be the case. Moreover, the atomic orbital model predicts that ρ will be large in the vicinity of each atom, and this is again verified by Fig. 6.11.

The atomic orbital model predicts that the charge distribution of X_{1c} will be similar to that of Γ_{1c}. However, as one can see by comparing Fig. 6.12 with Fig. 6.11, this is far from being the case. The small number of level contours for X_{1c} compared to the large number for Γ_{1c} shows that the charge density of X_{1c} is nearly constant in the unit cell, in sharp contrast to the situation at Γ_{1c} where strong peaks are centered on the atoms. The constancy of charge density associated with X_{1c} means that the atomic orbital model, although valid at $k = 0 = \Gamma$, is not valid elsewhere in momentum space for conduction band states.

But what, one may ask, is the significance of the charge density of conduction band states, since these are not occupied by electrons in the ground state of the crystal? In any polarization process the conduction band states will be involved, and they will be mixed into the valence band states by changes in chemical composition (e.g., the antisymmetric potential $V_A = V_{Ga} - V_{As}$ mixes the conduction band states of $V_S = (V_{Ga} + V_{As})/2 \approx V_{Ge}$ into the valence band states). The same is true of the effects of pressure and temperature. Moreover, when carriers are present in the conduction band, either by thermal activation or through doping with impurities, their interactions with the atoms of the lattice will differ greatly, according as to whether they occupy the Γ_{1c}, L_{1c}, or X_{1c} minima. Some of these differences are illustrated explicitly by the effects of pressure and temperature, which we now discuss.

PRESSURE DEPENDENCE OF BAND EDGES

Under hydrostatic pressure one finds substantial variations in the shifts in energy of various conduction band-valence band energy differences. Thus with hydrostatic pressure one can induce shifts in energy gaps similar to those that can be obtained by changing from one semiconductor compound to another, or by alloying one semiconductor with another. On the other hand, the application of uniaxial strain changes the overall crystal symmetry of diamond and sphalerite crystals from cubic to tetragonal [if the strain is parallel to a (100) cube axis] or rhombohedral [if the strain is parallel to a (111) body diagonal]. These changes in symmetry may cause splittings of states that were orbitally degenerate with cubic symmetry. Quite a lot of experimental information has accumulated concerning hydrostatic pressure dependences, while the data for uniaxial splittings are less complete and somewhat less accurate.

The hydrostatic results [4] of most interest concern the direct gap at Γ connecting the top of the valence band with $\Gamma_{2'c}$ (diamond symmetry) or Γ_{1c} (sphalerite symmetry), as well as the indirect gaps from the top of the valence band to L_{1c} and X_{1c}. The conventional notation for these three gaps is E_0, $E_{\Gamma L}$ and $E_{\Gamma X}$, respectively. Broadly speaking, the data show that dE_0/dP is about two to four times $dE_{\Gamma L}/dP$, both coefficients being positive, with dE_0/dP of order 10^{-5} eV/bar. The coefficient $dE_{\Gamma X}/dP$, on the other hand, is about 10^{-6} eV/bar, and is negative. These qualitative differences are readily understandable in terms of the different charge distributions in the unit cell of conduction band states as just discussed. Only the conduction band state at Γ is truly an antibonding state, the conduction band state at X being almost uniformly distributed, and that at L being intermediate between the other two. This is qualitatively the same as the ordering of the observed pressure coefficients.

To describe the observed coefficients quantitatively, use can be made of the scaling properties of the dielectric theory of energy gaps. These are described in more detail in the next chapter, but the general idea is that each gap scales with changes in bond length either because of pressure or changes in chemical composition. The scaling approach thus provides a connection between alloying and pressure experiments, in a way that seems consistent and satisfactory [4].

TEMPERATURE DEPENDENCE OF ENERGY GAPS

The observed temperature dependence of energy gaps in semiconductors nearly always consists of a reduction in the energy gap with increasing temperature. If we neglect the effects of thermal expansion (these can be removed using measured values of dE/dP), then the effects of atomic vibrations on energy levels can be calculated using either second-order perturbation theory or by modifying pseudopotential calculations in a manner analogous to the way in which Debye–Waller factors are introduced into the theory of X-ray scattering. The perturbation approach is simpler, and explains the qualitative effects immediately, while the dynamical pseudopotential theory [5] yields more satisfactory quantitative results.

In the perturbation picture we imagine that an electron at the bottom of the conduction band interacts with the thermal vibrations, and may be scattered to a state of higher energy in the same band. The rate at which this happens increases with the number of available phonons, and this in turn increases with temperature. According to second-order perturbation theory, when two states interact in this manner, they repel each other, the one of higher energy being pushed upwards, while the one of lower energy is pushed downwards. The energy of an electron at the bottom of the conduction band is therefore pushed downwards because of interactions with states of higher energy in the conduction band. Similarly, the energy of an electron at the top of the valence band is increased because of interactions with states of lower energy in the valence band. The combination of these two effects acts to produce the general reduction of the energy gap with increasing temperature. This is illustrated for $E_{\Gamma L}$ in Ge in Fig. 6.13. When the data are precise enough, it is possible to analyze the energies of the phonons involved, and to identify which are the strongly interacting states responsible for the observed shifts.

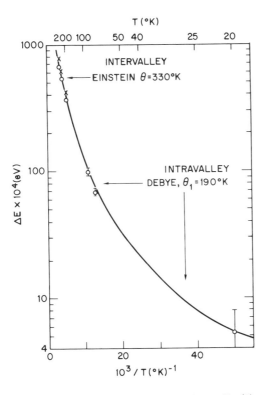

Fig. 6.13. Temperature dependence of the gap $\Delta E_{\Gamma L} = E_{\Gamma L}(0) - E_{\Gamma L}(T)$ in Ge. The theoretical curve contains two contributions, one corresponding to intravalley, which dominates at low temperatures, and one corresponding to intervalley, which dominates at high temperatures [from M. L. Cohen, *Phys. Rev.* **128**, 131 (1962)].

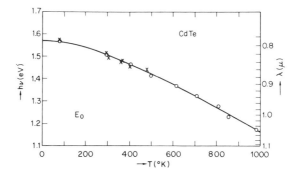

Fig. 6.14. Temperature dependence of the E_0 gap in CdTe [from D. de Nobel, *Philips Res. Rep.* **14**, 394 (1959)].

At high temperatures, i.e., well above the Debye temperature of the crystal, both acoustic and optic phonons contribute to the temperature dependence that is observed. Often measurements at room temperature and 78°K are used to define a "coefficient of linear temperature dependence" for a given band edge. Strictly speaking, such a linear temperature dependence is valid only well above the Debye temperature, and is not applicable near 78°K. Some idea of the differences between low-temperature and high-temperature behavior of an energy gap can be gained from studying Fig. 6.14, which shows the variation of the direct energy gap of CdTe over a wide temperature range. The value of dE_0/dT at high temperatures is clearly much greater than at or below room temperature.

SUMMARY

The covalent bond in semiconductors is characterized by accumulation of charge between nearest neighbors, which produces a well-localized bond charge. With increasing ionicity the bond charge is displaced towards the anion. When the bond charge is centered on the anion, the covalent structure becomes unstable and a first-order phase transition to the NaCl structure takes place. Pressure and temperature dependence of band edges and energy gaps can be calculated and explained by the quantum-mechanical methods used to obtain valence electron charge densities.

REFERENCES

1. J. C. Phillips, "Covalent Bonding in Crystals, Molecules and Polymers." Univ. of Chicago Press, Chicago, Illinois, 1970.
2. M. L. Cohen and V. Heine, *Solid State Phys.* **24** (1970).

3. J. C. Phillips, *Rev. Mod. Phys.* **42,** 317 (1970).
4. D. L. Camphausen, G. A. Neville Connell, and W. Paul, *Phys. Rev. Lett.* **26,** 184 (1971).
5. P. Y. Yu and M. Cardona, *Phys. Rev.* **2B,** 3193 (1970); J. P. Walter, *et al.*, *Phys. Rev. Lett.* **24,** 102 (1970).

7

Fundamental Optical Spectra

Almost all our knowledge of the electronic structure of atoms comes from spectroscopic studies of line spectra. Similarly by far the greatest proportion of our knowledge of the electronic structure of semiconductors stems from theoretical analysis of the optical reflectivity from clean crystalline surfaces. While the spectral structure of atoms was systematized in the nineteenth century and its principal features explained before 1920 by the semi-classical Bohr model of the atom, systematic understanding of semiconductor spectra began in 1962. Emphasis in this discussion is placed on these recently understood general features, as the reader who desires a more technical discussion may choose between several longer treatments [1-3].

The part of the optical spectrum of a perfect crystal which is independent of impurities or defects is called intrinsic, as distinguished from extrinsic absorption caused by impurities or by lattice defects, such as vacancies. Suppose we further restrict our attention to direct electronic transitions in which the crystal momentum $\mathbf{p} = \hbar\mathbf{k}$ of the electrons is conserved, thereby omitting from consideration the indirect transitions which change $\mathbf{k}$ through absorption or emission of a phonon. This direct part of the spectrum is called fundamental, because it is characteristic of the pure crystal with the atoms at rest, apart from zero-point vibrations, whose effect in semiconductors is negligible in most cases.

The fundamental optical spectrum of a semiconductor is best studied at low temperatures, but it is not greatly altered at room temperature, although some broadening of the structure is observable. The spectral range covered by the fundamental optical absorption of most semiconductors starts at quite low energies (comparable to ΔE_{cv}), peaks near E_g, and then tails off to smaller values by the time $\omega = \omega_p$, the plasma frequency. See Table 2.1 for some representative energies.

ONE-ELECTRON EXCITATIONS

Almost all of the transitions that contribute to optical absorption in atoms, molecules, or crystals can be described as one-electron excitations. The chief reason for this in atoms is that the strongest term in the potential seen by a valence electron consists of the Coulomb potential of the ion core, of order Ze/r_a, where r_a is the atomic radius. The interaction with another valence electron is of order only $0.3e/r_a$, so that the probability is small that because of their interaction the first electron will be excited together with the second.

In crystals collective excitations of many electrons (plasma excitations) may be possible, but again in most semiconductors their contribution does not begin to be important until the energies that are involved are of order $\hbar\omega_p$, which is much larger than E_g. We are therefore concerned with one-electron excitations, which may be of two kinds.

LINE AND CONTINUUM (BAND) SPECTRA

Atomic valence spectra are of two types. There are series of line spectra, corresponding to excitations of valence electrons to larger bound orbitals. Each series approaches a limit corresponding to the ionization energy of the valence level in question. Beyond the series limit there is continuum absorption corresponding to excitation of the valence electron into a state of positive energy above the energy of an electron in vacuum at rest. Most of the absorption or oscillator strength is associated with the line spectrum (transitions to states where the excited electron is still bound to the positively charged ion), and relatively little is left for the continuum absorption.

The fundamental optical spectra of semiconductors are also of two types.

Denote by E_0 the smallest energy for which a direct transition from the valence band to the conduction band is possible. For frequencies less than $\omega_0 = E_0/\hbar$, line spectra are observed. These correspond to an electron in a superposition of conduction-band states bound to a missing electron (a positively charged hole) in a superposition of valence-band states. Such bound states are called excitons, and they are discussed in detail in a later chapter.

For frequencies $\omega > \omega_0$ the fundamental optical spectrum is continuous, corresponding to excitations of free (unbound) electrons and holes. The two kinds of spectra which are found in atoms, line and continuum, are therefore found in semiconductors as well. However, as indicated in Fig. 7.1, there is an important quantitative difference between the two cases. While the line spectra of atoms contain most of the oscillator strength associated with valence-electron excitation, in semiconductors line spectra are typically hundreds of times weaker than in atoms. Almost all the oscillator strength of the valence electrons in semiconductors is associated with continuum absorption between valence and conduction bands. This is called direct interband absorption, and it is the subject of the remainder of this chapter.

DIELECTRIC FUNCTION

The optical absorption of solids can be described using either the complex index of refraction $n + ik$ or the complex dielectric function $\epsilon_1 + i\epsilon_2$. Here n, k, ϵ_1 and ϵ_2 are all real functions of the frequency ω and they are related at any given frequency according to

$$\epsilon_1 = n^2 - k^2, \qquad \epsilon_2 = 2nk. \tag{7.1}$$

We saw earlier that the one-electron approximation is an appropriate one for discussing optical transitions in semiconductors. To this approximation another one may be added, namely, that the local electric field causing optical transitions in the crystal is uniform and therefore equal to the macroscopic electric field of the incident photons. In metals where the valence electrons are almost uniformly distributed outside the atomic cores, neglect of local field corrections is well justified. In semiconductors the situation is not so clear cut, for as we saw in the preceding chapter there is appreciable charge localization in valence-band states and some conduction-band states. In practice theoretical calculations of optical spectra which have been made without local field corrections have yielded in most cases results which agree with observation to within the limits of experimental uncertainties.

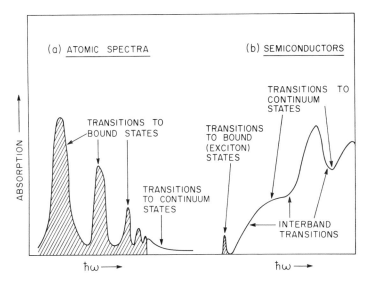

Fig. 7.1. Comparison of relative strengths of line and continuum spectra in atoms and semiconductors. The widths of the atomic spectral lines have been exaggerated in order to emphasize their great oscillator strength relative to continuum absorption.

With these simplifications one may derive a formula for $\epsilon_2(\omega)$ including only direct valence-conduction interband transitions,

$$\epsilon_2(\omega) = \frac{\alpha}{\omega} \sum_{v,c,k} \frac{\langle c, \mathbf{k} \mid \mathbf{e \cdot p} \mid v, \mathbf{k} \rangle^2}{E_c(\mathbf{k}) - E_v(\mathbf{k})} \delta(E_c - E_v - \hbar\omega), \qquad (7.2)$$

where α is a constant, $\mathbf{e}$ is the polarization vector of the incident light, $\mathbf{p}$ is the electron momentum, and the δ function describes the conservation of energy.

SUM RULES

The functions $\epsilon_1(\omega)$ and $\epsilon_2(\omega)$ are connected by a general relation which is valid whether or not one makes the one-electron approximation and neglects local field corrections, as one does in deriving (7.2). Thus the Kramers–Kronig relation

$$\epsilon_1(\omega) = 1 + \pi^{-1} \int_{-\infty}^{\infty} \frac{\epsilon_2(\omega')}{\omega' - \omega} d\omega' \qquad (7.3)$$

is always valid [1]. A second general relation connects the magnitude of

$\epsilon_2(\omega)$ with the total electron density of the crystal. This relation is

$$\int_0^\infty \omega\epsilon_2(\omega) \ d\omega = \pi\Omega_p^2/2, \tag{7.4}$$

where Ω_p^2 involves the total number of electrons N_T per unit volume V of the crystal,

$$\Omega_p^2 = 4\pi N_T e^2/mV. \tag{7.5}$$

Of course, most of the electrons in the crystal are core electrons which are tightly bound to the atomic nuclei. Let ω_c be a cutoff frequency which is large compared to $E_g/\hbar$ (where E_g is of order 1 Ry) and still small compared to core excitation frequencies $E_c/\hbar$ (where E_c is of order ten Ry). Then one can replace (7.4) with a relation which is not exact but which is much more useful,

$$\int_0^{\omega_c} \omega\epsilon_2(\omega) \ d\omega = \pi\omega_p^2/2, \tag{7.6}$$

where ω_p^2 is now defined in terms of the effective number of valence electrons N per unit volume V,

$$\omega_p^2 = 4\pi N e^2/mV. \tag{7.7}$$

Thus ω_p is the valence electron plasma frequency [compare to Eq. (2.5)].

DIRECT THRESHOLDS

Because $E_c(\mathbf{k})$ and $E_v(\mathbf{k})$ describe energy bands rather than discrete energy levels, it might appear that the optical spectrum described by (7.2) should be broad and featureless. The spectrum is indeed continuous, but it has quite definite analytic features that have proved to be of great value in comparing experimental results to theoretical models. These features can be illustrated by evaluating (7.2) near a direct threshold.

To simplify the calculation let us suppose the conduction band and valence band are nondegenerate and have the form shown in Fig. 7.2. Near $\mathbf{k} = 0$ the valence band energy is given by

$$E_v(\mathbf{k}) = E_v(0) - \hbar^2 k^2/2m_v \tag{7.8}$$

and the conduction band energy is given by

$$E_c(\mathbf{k}) = E_c(0) + \hbar^2 k^2/2m_c. \tag{7.9}$$

The energy E_0 of the direct interband threshold is given by

$$E_0 = E_c(0) - E_v(0). \tag{7.10}$$

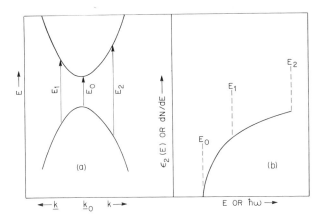

Fig. 7.2. Two-band model for a direct optical threshold. Some of the allowed, **k**-conserving interband transitions are indicated by arrows in Fig. 7.2a. In Fig. 7.2b the corresponding optical spectrum is shown.

The problem is to evaluate (7.2) over the range of energy near E_0 where the parabolic approximations (7.8) and (7.9) are valid. In general $\epsilon(\omega)$ is a tensorial function, because it connects the displacement vector **D** with the electric field **F**. In a cubic crystal, however, the tensor becomes a scalar because of symmetry $[\epsilon_{ij}(\omega) = \epsilon(\omega)\delta_{ij}]$. This point is implicit in the matrix elements in (7.2), which we take to be constant and independent of the polarization vector **e**. With this further simplification near E_0 Eq. (7.2) simplifies to

$$\epsilon_2(\omega) = \frac{\text{const}}{\omega^2} \sum_{\mathbf{k}} \delta[E_c(\mathbf{k}) - E_v(\mathbf{k}) - \hbar\omega]. \qquad (7.11)$$

The region in **k** space where the argument of the δ function in (7.11) vanishes is the surface of a sphere centered on $\mathbf{k} = 0$ with radius q given by

$$[\hbar^2/2m_v + \hbar^2/2m_c]q^2 = \hbar\omega - E_0 \qquad (7.12)$$

for $\hbar\omega > E_0$. For $\hbar\omega < E_0$, of course, (7.11) vanishes. The δ function in (7.11) asks us to evaluate the number of states N per unit energy $E = E_c - E_v$,

$$dN/dE = (dN/dq)[1/(\partial E/\partial k)]. \qquad (7.13)$$

The quantity dN/dq is the area of the spherical surface, which is proportional to q^2. The second factor in (7.13) is proportional to $[(m_v^{-1} + m_c^{-1})q]^{-1}$, so

$$\omega^2\epsilon_2(\omega) \propto dN/dE \propto m_v m_c q/(m_v + m_c)$$
$$\propto [m_v m_c/(m_v + m_c)]^{1/2}(E - E_0)^{1/2}. \qquad (7.14)$$

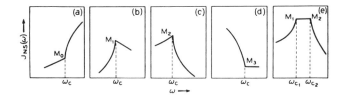

Fig. 7.3. Analytic features of continuum band absorption in crystals: (a) threshold contribution, (b), (c) saddle-point edges of two kinds, (d) cutoff absorption at interband maximum, (e) confluence or quasi-degeneracy of saddle edges of two kinds.

The optical spectrum described by (7.14) is also plotted in Fig. 7.3a. In one or two dimensions dN/dq is proportional to one or to q, respectively. Then $\epsilon_2(\omega)$ is proportional to $(E - E_0)^{-1/2}$ or to a constant, respectively; i.e., the direct threshold in two dimensions makes $\epsilon_2(\omega)$ a step function, which is zero for $\hbar\omega < E_0$ and constant (to lowest order) for $\hbar\omega > E_0$. In one dimension $\epsilon_2(\omega)$ is singular like $(E - E_0)^{-1/2}$ as $E \to E_0$ from above.

The special cases that have been discussed here contain most of the elements which are present in the more general case. The function

$$E(\mathbf{k}) = E_c(\mathbf{k}) - E_v(\mathbf{k}) \tag{7.15}$$

will in general have a minimum at some point $\mathbf{k} = \mathbf{k}_0$ in the Brillouin zone. At this point $\nabla_k E(\mathbf{k}) = 0$. Any point satisfying this extremal condition, even if it is not a minimum, is called a *critical point*. About such a point the integral (7.2) can be studied to second-order in $\mathbf{k} - \mathbf{k}_0$. In three dimensions all such critical points generate edges in dN/dE which contain terms with square root singularities similar to the one contained in (7.14). This is true not only when the critical point is a minimum or maximum of $E(\mathbf{k})$, but also when it is a saddle point. Near a saddle point of energy E_1, for example, $\epsilon_2(\omega)$ might be described by

$$E \leq E_1, \qquad \epsilon_2(\omega) = A - B(E_1 - E)^{1/2}, \tag{7.16a}$$

$$E \geq E_1, \qquad \epsilon_2(\omega) = A + C(E - E_1), \tag{7.16b}$$

where A, B, and C are constants. Most of the observed singularities behave either like (7.14) or (7.16), as shown in Figs. 7.3a–7.3d. More complicated behavior can be encountered when two singularities are accidentally degenerate to within an energy comparable to that of broadening effects, as shown in Fig. 7.3e.

GERMANIUM

As an example consider $\epsilon_2(\omega)$ for Ge as shown in Fig. 7.4. The experimental curve is based on reflectivity data obtained over an energy range wider than that shown in the figure. From the reflectivity one can obtain $\epsilon_2(\omega)$ by using a Kramers–Kronig relation similar to (but not the same as) Eq. (7.3). The experimental curve shown in Fig. 7.4 exhibits the gross features of the spectrum. These are: (1) The direct threshold at $E_0 = 0.8$ eV, (2) a steep edge near $E_1 = 2.1$ eV, and (3) a large peak near $E_2 = 4.4$ eV. Other features are indicated in the figure, but these three illustrate the influence of critical points on the spectral structure of continuum absorption in predominantly covalent semiconductors such as silicon and germanium.

The direct threshold at E_0 is associated with transitions near $\Gamma_{25'} \to \Gamma_{2'}$ (see Fig. 5.12). According to (7.14), its strength is proportional to $m_c^{1/2}$ for the case $m_v \gg m_c$. This is the case for the Γ effective masses, and in Ge, according to Table 5.1, the value of m_c/m is only 0.038. This makes the E_0 edge weak, as one could also infer qualitatively from the energy bands of Ge, shown in Fig. 5.12. There we see that the volume of k space per unit energy occupied by energy levels in the conduction band near $\Gamma_{2'}$, is small.

The E_1 edge is about 50 times stronger than the E_0 edge. The theoretical curve shown in Fig. 7.4 was calculated from the energy bands of Ge (shown in Fig. 5.12) and wave functions obtained from the pseudopotential form

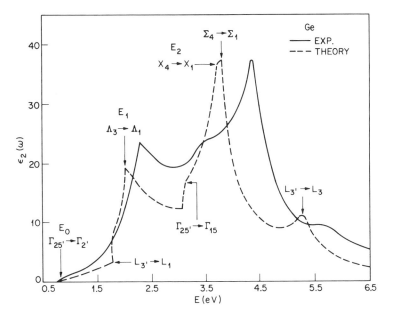

Fig. 7.4. The optical spectrum of Ge. The solid curve is experimental, while the dashed curve is based on theory. Important interband edges are indicated, and can be identified with **k**-conserving transitions in the regions of the energy bands of Ge shown in Fig. 5.12.

factor shown in Fig. 6.3. Because the theoretical curve agrees so well with the experimental one, one can examine the calculated energy bands to identify the location and nature of the critical points in $E(\mathbf{k}) = E_5(\mathbf{k}) - E_4(\mathbf{k})$ where $\nabla_\mathbf{k} E = 0$. These critical points must be responsible for the steepness and magnitude of the E_1 edge. It turns out that there is a threshold [Fig. 7.3b] near 2.1 eV of the type described by equations (7.16). This makes the edge steep, and it rises to a high value near 2.1 eV because the appropriate conduction and valence band masses are much larger than the masses for the E_0 threshold. Finally, one can ascertain the locations of the critical points responsible for the subsidiary threshold and the saddle point. These are at $\mathbf{k} = \mathbf{L}$ and along the $\mathbf{k} = \Lambda$ line, and are indicated by M_0 and M_1 transitions in Fig. 5.12.

The explanation for the E_2 peak, where $\epsilon_2(\omega)$ reaches its highest value, can also be given in terms of critical points. The structure begins with a threshold at 4.0 eV, rises to a saddle point edge of the type shown in Fig. 7.3b at 4.3 eV, and falls off rapidly because of a second-saddle point edge [Fig. 7.3c, the reverse of (7.16)] at 4.4 eV. Thus the E_2 peak is caused by an accidental confluence of two saddle point edges (Fig. 7.3e), which curiously enough is found in most $A^N B^{8-N}$ crystals. The critical points involved are located near the (200) and (110) edges of the Brillouin zone, far from the regions contributing to the E_0 and E_1 edges.

PHOTOEMISSION

When an electron is excited from a state in the valence band to a state in the conduction band by absorbing a photon, it may escape from the crystal if its energy in the final state lies above the vacuum level. The yield of electrons so escaping can be counted, and their energy above the vacuum level can be measured by applying a suitable retarding potential. In an ordinary optical experiment at a given frequency ω only one or two numbers [such as the reflectivity $R(\omega)$] are measured. Here for a given frequency ω an entire distribution of electron energies, $dN_\omega(E)/dE$, can be determined. In this sense photoemission contains more information than optical measurements. However, photoemission suffers from several disadvantages. The vacuum level may lie well above most of the conduction band states near the energy gap, and of course it is these states which are of greatest interest. In addition most of the electrons which escape from the crystal without inelastic scattering were excited quite close ($\lesssim 25$ Å) to the surface, so that surface preparation, which was already critical for ordinary optical measurements, becomes even more critical in photoemission experiments. Finally, the resolution of the energy distribution using a retarding potential may not be much better than 0.1 eV, compared to resolution of order 0.002 eV achievable by the derivative techniques discussed in the next section.

The photoemission yield $Y_W(\omega)$ measures the number of electrons emitted per photon absorbed as a function of photon frequency ω. It is dependent on the work function W of the crystal, which is defined as the position of the vacuum level relative to the top of the valence band at the surface of the crystal. By evaporating a fraction of an atomic layer of Cs on a cleaned surface, W can be varied from a typical value of 5 eV to less than 2 eV. Microscopically, $Y_W(E)$ is given approximately by

$$Y_W(\omega) = \sum_{i,j} f_{ij} N(E_{ij}) p(i) / \sum f_{ij} N(E_{ij}), \qquad (7.17)$$

where f_{ij} is the oscillator strength, $E_{ij} = \hbar\omega$, $N(E_{ij})$ is the interband density of states, and $p(i)$ is the probability that an electron in the ith conduction band state will escape from the crystal. As W is reduced by cesiating the surface, $p(i)$ changes, as conduction band states of lower energy are progressively uncovered.

As an example of photoemission data, the energy distribution of the photoelectric yield from Ge covered with a monolayer of Cs is shown in Fig. 7.5. In this case the value of W is 2.35 eV. Referring to Fig. 5.12, we see that this places almost the entire first conduction band below the minimum energy necessary for an electron to escape from the crystal, while most of the states associated with higher conduction bands do have sufficient energy to escape from the crystal. Thus transitions to final states near Γ_{15} and L_3 are expected to give peaks in Y. In Fig. 5.12 these transitions are labeled as $E_0' = 3.2$ eV and $E_1' = 5.4$ eV, respectively, and they are associated with transitions labeled $\Gamma_{25'} \rightarrow \Gamma_{15}$ and $L_{3'} \rightarrow L_3$, respectively. Peaks in $Y(E)$ at the appropriate energies are indeed seen in Fig. 7.5. On the other hand, transitions near X labeled $E_2 = 4.4$ eV in Fig. 5.12 correspond to final states near X_1, which is below vacuum level. Hence these transitions produce a dip in $Y(E)$, which is seen at the appropriate energy in Fig. 7.5 and is labeled $X_4 \rightarrow X_1$. Direct comparison is made between this dip and the corresponding peak in $\epsilon_2(\omega)$ in Fig. 7.5, and it is seen that the agreement is very good. Similarly good agreement is found in other crystals.

Transitions near $\Gamma_{25'} \rightarrow \Gamma_{15}$ have nearly the same energy over a small volume of **k** space, and hence give rise only to a weak knee in $\epsilon_2(\omega)$ near 3.5 eV in Ge, as shown by the lower part of Fig. 7.5. The initial states, however, at $\Gamma_{25'}$ are the ones of highest energy in the valence band, and this means that photoelectrons excited from the neighborhood of $\Gamma_{25'}$ have a much greater probability $p(i)$ of escaping from the crystal than those excited from other lower parts of the valence band by photons in the energy

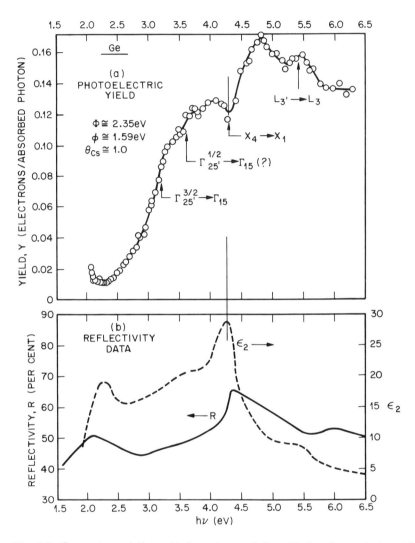

Fig. 7.5. Comparison of the optical spectrum of Ge with its photoemissive yield spectrum $Y(\omega)$. The sample surface was prepared by cleaving the crystal in vacuum and then covering the freshly cleaved surface with one monolayer of Cs.

interval near 3.5 eV. In fact the triply-degenerate p-like states $\Gamma_{25'}$ (see Fig. 5.10a) are spin–orbit split by 0.3 eV in Ge. (Spin–orbit splittings are discussed in more detail later in this chapter). This splitting is difficult to resolve in $\epsilon_2(\omega)$, but it is quite prominent in the yield curve shown in Fig. 7.5. The two spin–orbit split levels are labeled with $J = \frac{3}{2}$ and $J = \frac{1}{2}$, respectively, and are so labeled in Fig. 7.5.

The minimum direct gap in Si probably occurs very near Γ, and may be labelled $\Gamma_{25'} \rightarrow \Gamma_{15}$. (In this case the spin–orbit splitting is less than 0.05 eV and so is not resolved). A theoretical yield curve based on the pseudo-potential energy bands of Fig. 5.11 and using equation (7.17) to calculate $Y(\omega)$ is compared with the experimental data for Si covered with a mono-layer of Cs in Fig. 7.6. The overall agreement on peaks and dips is quite good; the small shifts for $X_4 \rightarrow X_1$ and $L_{3'} \rightarrow L_3$ come from the fact that

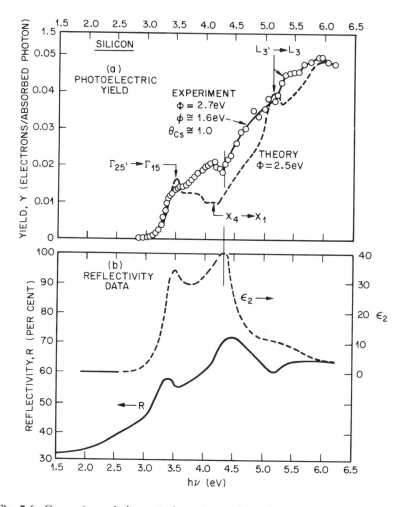

Fig. 7.6. Comparison of the optical spectrum of Si with its photoemissive yield spectrum Y(ω). Surface preparation as described for Fig. 7.5. Experimental data of Fig. 7.5 and Fig. 7.6 by F. Allen and G. Gobeli, quoted and analyzed in M. L. Cohen and J. C. Phillips, *Phys. Rev.* **139,** A512 (1965).

the pseudopotential parameters were overdetermined, and so could not fit these energies exactly. Of particular interest is the very good fit to the threshold in $Y(\omega)$ in the region between 3.0 eV and 3.5 eV, based on $E_0' = E(\Gamma_{15}) - E(\Gamma_{25'}) = 3.4$ eV. A number of so-called "first principles" or "*ab initio*" one-electron energy band calculations have consistently given smaller values for this quantity, in the range 2.4 ∼ 2.8 eV. This has led some band theorists to question the experimental value of E_0', and to claim (as much as ten years after the experimental data were published and analyzed correctly!) that E_0' should be near 2.5 eV. In all likelihood, the difficulty here lies in the one-electron approximation itself, since there is no possibility for an error of much more than 0.1 ∼ 0.2 eV in the determination of this energy from the experimental data shown in Fig. 7.6.

DERIVATIVE TECHNIQUES

In addition to the gross features of optical spectra just discussed, there is a good deal of fine structure to be uncovered [3]. In principle some of this can be obtained by very accurate studies of reflectivity, but in practice higher resolution and greater reliability have been achieved by modulating the reflectivity $R(\omega)$ with respect to some parameter α. The modulated reflectivity may be represented by $dR/d\alpha$. The variables used for α have been ω itself (which has the advantage of not disturbing the crystal), the temperature T, a uniaxial strain field ϵ_{xx}, and an electric field F_x applied both normal to a yz crystal face or parallel (F_y) to a yz one. The last two techniques yield a great deal of information, especially when used in conjunction with linearly polarized light.

In all cases studied so far almost all the structure in $dR(\omega)/d\alpha$ occurs for $\hbar\omega$ equal to an interband energy $E(\mathbf{k})$ for which $\nabla_k E(\mathbf{k}) = 0$, that is, the derivative structure is associated with critical points. The structure most commonly observed by derivative methods, which is difficult to resolve in direct reflectivity, arises from spin–orbit splitting of interband edges. This is evident in Fig. 7.7, which shows dR/dF_x for Ge. The spin orbit splittings of the E_0 and E_1 peaks (denoted by $E_0 + \Delta_0$ and $E_1 + \Delta_1$, respectively) are easily resolved. In Ge and almost all $A^N B^{8-N}$ tetrahedrally coordinated semiconductors, these peaks have been resolved. Moreover in general $\Delta_1 = 2\Delta_0/3$, a result which can be explained in terms of the threefold orbital degeneracy of the levels at Γ responsible for the E_0 edge compared to the twofold orbital degeneracy of the levels along Λ responsible for the E_1 edge.

The interpretation of derivative spectra, such as that obtained by the electroreflectance technique, is complicated by the changes in the band structure of the crystals induced by the modulation field. For example, the

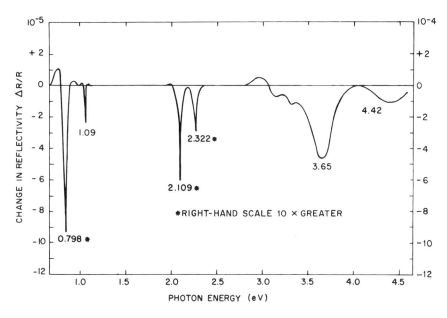

Fig. 7.7. The reflectance spectrum of Ge as modulated by an electric field F_x. Shown here is $R(F_x,\omega) - R(0,\omega)$ where R is the reflectivity of the crystal [from B. O. Seraphin and R. B. Hess, *Phys. Rev. Lett.* **14**, 138 (1965)].

modulating electric field induces tunneling by electron wave packets localized (in **k** space) near band edges, and it is this tunneling that changes the reflectivity. If one wishes to make a very simple derivative measurement the easiest thing to interpret is the reflectivity modulated with respect to the wavelength λ (or frequency ω) of the incident light. Plots of $d\log R(\omega)/d\omega$ for Si and for GaAs are shown in Fig. 7.8 and Fig. 7.9, where they are compared directly with theoretical pseudopotential energy band calculations. The overall agreement is very good, although some of the theory and experiments differ in fine structure, corresponding to band shifts of order 0.1 eV.

INTERBAND ENERGIES

In this way enough interband edges have been resolved to establish the positions of the valence and conduction bands of these $A^N B^{8-N}$ compounds to a degree of completeness unparalleled in any other family of solids. Most of the interband energies shown in the figures of Chapter 5 were obtained in this way. All interband energies, both direct and indirect, which are

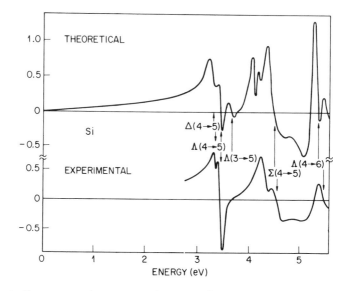

Fig. 7.8. Comparison of experimental and theoretical values of $d \log R(\omega)/d\omega$ in Si. The theoretical values are based upon pseudopotential energy bands of the kind described in Chapter 5. The locations of the neighborhoods in **k**-space giving rise to the oscillations in the theoretical spectrum are identified in terms of conventional symmetry labels of the valence and conduction bands involved in the corresponding optical transitions.

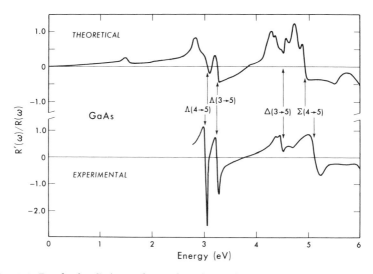

Fig. 7.9. Results for GaAs similar to those for Si shown in Fig. 7.8. The experimental data and theoretical calculations are reported by R. R. L. Zucca and Y. R. Shen, *Phys. Rev. B* **1**, 2668 (1970).

TABLE 7.1

Direct and Indirect Interband Energies in 18 Semiconductors (eV)[a]

Crystal	I	E_X	E_L	E_0	E_1	E_{2A}	E_{2B}	E_0'	E_1'
C		5.48				12.2	12.2	7.3	
BP		2.0			5	6.9			8.0
SiC		2.3		7.75	7.1	8.3		6.0	9.7
Si	5.17	1.13		4.18	3.45	4.40	4.40	3.35	5.4
GaP		2.38	2.5	2.77	3.73	5.27	5.74	4.80	6.6
ZnS	8.7			3.80	5.70	7.6	8.75	6.65	9.45
Ge	4.90	0.96	0.76	0.89	2.26	4.3	4.3	3.32	5.4
AlSb	5.47	1.87		2.5	3.0	4.25	4.6	3.8	5.1
GaAs	5.59	1.92	2.0	1.55	3.04	4.85	5.33	4.65	6.2
InP	5.72	2.1	2.0	1.37	3.24	4.8	5.1	4.4	6.6
ZnSe	7.55			2.68	4.95	6.4	7.2		8.45
CdS	7.35			2.48	5.3		7.8	6.4	8.8
GaSb	5.03	1.30	1.07	0.99	2.26	4.1	4.5	4.0	5.5
InAs	5.44			0.5	2.64	4.5	5.0	4.5	6.1
ZnTe	5.89			2.56	3.97	5.3		5.40	7.20
CdSe	6.88			1.87	4.5		6.9	6	8.35
InSb	5.07			0.5	2.13	4.08		3.8	5.25
CdTe	6.01			1.80	3.69	5.00	5.9	5.07	6.8

[a] Here I is the ionization potential, E_X the minimum indirect band gap from the valence band maximum at Γ to the conduction band minimum at or near X, E_L the similar indirect gap to L, while the meaning of the direct gaps $E_0, \ldots, E_1'$ is discussed in the text. See also Fig. 7.4 for the case of Ge.

known to the author are cataloged in Table 7.1 for 18 common semiconductors.

CORE d ELECTRONS

The lattice constants of Si and Ge differ by less than 4%, but the energy bands of these two crystals (see Fig. 5.11 and Fig. 5.12) differ very much, especially as regards the E_0 and E_1 energy gaps. Indeed comparison of the energy bands of the two crystals shows that the conduction band connecting $\Gamma_{2'}$ with L_1 for **k** along (111) directions has dropped about 2 eV on going from Si to Ge, with the largest drop being associated with $\Gamma_{2'}$ (about 3 eV) and the smallest with L_1 (about 1.5 eV).

The explanation for this behavior is based on the distribution of probability density of the $\Gamma_{2'}$ and L_1 states in the unit cell. As we saw in the last chapter, these two states are not typical of conduction-band states because they are centered on the two atoms in the unit cell rather than being more uniformly distributed. Moreover, because of their s-like character they are able to penetrate the 3d shell (in the case of 4s valence electrons in Ge), which lowers their energy considerably. This is shown even in the free atom by plotting the ratio of np to ns valence-binding energies as a function of n (see Fig. 1.8). As n increases both $| E_{np} |$ and $| E_{ns} |$ decrease, because of increasing atomic size. However, $| E_{ns} |$ does not decrease as fast as $| E_{np} |$, because the ns valence electrons can penetrate the increasingly large atomic core. Notice that for the atoms from the C, Si, and Ge rows this trend is produced by the addition of p electrons to the core (in the case of Si compared to C) and by the addition of d electrons to the core (in the case of Ge compared to Si). For atoms from the Sn and Pb rows, saturation would occur were it not for the increasing influence of relativistic effects quite near the nucleus, as also evidenced by the relativistic increase in binding of 1s electrons (not shown in Fig. 1.8).

While these effects are noticeable in free atom calculations, they are even more drastic in crystals. In Ge, for example, the threshold for excitation of 3d core electrons to the conduction band begins around 30 eV. Thresholds for A cations (such as Ga or In, see Fig. 7.10) start at lower energies. However, the mixing of valence and conduction band states separated by energies of 30 eV is influenced by the presence of the core d states. This point can be brought out by using Eq. (7.6) to evaluate ω_p^2 and thereby determine N, the effective number of valence electrons given by Eq. (7.7). When this is done for Si, N turns out to be four per atom, just as one would expect from conventional chemical models. A plot of N as a function of the cutoff frequency ω_c is shown in Fig. 7.11 for several semiconductors. For Si there is a rapid rise in $N(\omega_c)$ near $\hbar\omega_c = E_h \approx 5$ eV followed by a trend towards saturation in the region 10–20 eV at a value of N nearly equal to four per atom. The remaining semiconductors, however, all contain at least one element with $(n - 1)$d core electrons which interact appreciably with the ns and np valence electrons. As a consequence, after similar rapid rises in $N(\omega_c)$ for $\hbar\omega_c \approx E_g$, the value of N approaches saturation more slowly and moreover it can be seen that the saturation value will be considerably greater than four. In the case of InSb, for example, the extrapolation shown in Fig. 7.11 suggests that N may level off near a value of 5.7 per atom.

While the sum rule in Eq. (7.6) is of interest in its own right, the application just discussed is important for several other reasons. First, and most important, it shows that there is an operational spectroscopic procedure which can be employed to analyze quantitatively the effect of core d states on s and p valence electron states. If we were to attempt to analyze this effect by solving the Schrödinger equation in the crystal and studying the

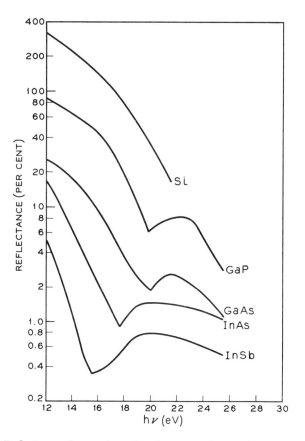

Fig. 7.10. Reflectance of several semiconductors at intermediate energies. Starting from 12 eV the reflectance decreases, representing the exhaustion of bonding → anti-bonding oscillator strength at energies greater than $2E_g$. The rise in reflectivity in the 15–20 eV range in the Ga and In compounds is caused by excitation of electrons from Ga cores (3d states) or In cores (4d states). See also Figs. 2.1 and 2.2.

resulting wave functions, the complexity of the latter would leave us at a loss as to how to proceed to discuss the effect of d states on bonding. Let us denote the saturation value of $N(\omega_c)$ by N and define D by

$$D = N/4, \qquad (7.18)$$

so that $D = 1.00$ for crystals containing no d electrons in atomic cores, while $D > 1$ when the atoms contain d electrons in their cores. We can use the parameter D as a quantitative measure of how much the s valence electrons penetrate the atomic core and therefore see an effective core

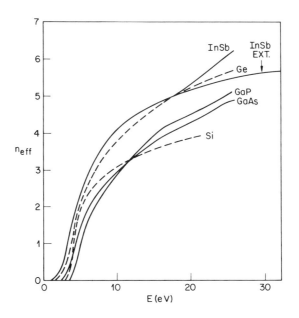

Fig. 7.11. The effective number of valence electrons, $N(\omega_c)$, as a function of cutoff frequency $E = \hbar\omega_c$ for several semiconductors. Note the qualitative difference between the curve for Si and that for the other semiconductors shown.

charge greater than what one would expect from conventional valence considerations.

Using the parameter D we can now state the precise form of Eq. (2.20), the relation used to define the average energy gap E_g between valence and conduction band or bonding and antibonding energy levels. This relation is based on the model band structure shown in Fig. 5.5 and it is

$$\epsilon(0) = 1 + (\hbar\omega_p/E_g)^2 D(1 - E_g/4E_F), \qquad (7.19)$$

where ω_p^2 now is defined in terms of the number N of conventional valence electrons per atom. The last factor in (7.19) allows for the spread in energy bands shown in Fig. 5.5, according to which the minimum energy difference between valence and conduction bands is E_g. Most of the optical transitions with large oscillator strength are between valence band states in Fig. 5.5 with k slightly less than k_F and $E_v(k)$ just slightly less than $E_F - E_g/2$. Thus the average excitation energy is somewhat more than E_g, which reduces $\epsilon(0)$ by the last factor in Eq. (7.19), as discussed in connection with Eqs. (5.12)–(5.14).

If we compare the optical spectrum of Ge, 7.4, with the optical spectrum of an isotropic semiconductor, Eq. (5.13), with $E_\mathrm{h} = 4.2$ eV $\simeq E_2$, we see that (5.13) describes $\epsilon_2(\omega)$ at the E_2 peak and above, while the E_0 and E_1 structures are represented by the presence of the factor $D > 1$. The connection between the E_0 and E_1 energies and D is described in a later section of this chapter (Eqs. 7.26 and 7.27).

SPECTROSCOPIC DEFINITIONS OF VALENCE

In chemical usage valence is a word which may mean several different things. Often it is used to label the column of the periodic table to which a cation A^N belongs ($N \leq 4$) or to specify the number of electrons $M \leq 4$ missing from the valence shell of an anion B^{8-M}. In a molecular context valence may specify the largest coordination number commonly found in complexes containing the atom in question. In most cases these definitions yield the same result which equates valence to the number of ns–np electrons present in or missing from the last partially filled (n) shell of the atom. In some cases, however, the d electrons in the filled $(n-1)$d shell are sufficiently weakly bound to contribute to the valence. This is the case, for example, with the IB group metals Cu, Ag, and Au. In molecules these atoms are often twofold coordinated, indicating that approximately one $(n-1)$d electron is forming bonds in addition to the one ns electron.

In the crystals CuX and AgX, where X is a halogen atom, it would appear that we are dealing with $A^N B^{8-N}$ compounds with $N = 1$. However, when we examine the experimental values of the ionic energy gap C, we find that these can be fitted with the formula previously discussed in Chapter 2,

$$C = 1.5(Z_\mathrm{A}e^2/r_\mathrm{A} - Z_\mathrm{B}e^2/r_\mathrm{B}) \exp(-k_s R), \qquad (7.20)$$

with $Z_\mathrm{A} = 2$ and $Z_\mathrm{B} = 7$. This means that one of the d electrons of the IB metal atom is indeed contributing to reducing the difference in electronegativity $X_\mathrm{A} - X_\mathrm{B}$ represented by the term in parentheses in Eq. (7.20).

This example, combined with the discussion of the increase in valence oscillator strength represented by the factor D, shows that spectroscopic discussion of the structural implications of bonding can go well beyond considerations of classical valence theory without resorting to detailed study of electronic wave functions. Moreover, as we shall now see, the chemical analysis of trends in bond energies in terms of trends of the *average* energy gap E_g can be applied equally well to the principal interband energies, E_0, E_1, E_2, etc., associated with transitions in various local regions of the Brillouin zone.

CHEMICAL TRENDS IN INTERBAND ENERGIES

Direct interband transitions which do *not* involve the s-like conduction band states on or near the Γ–L line in **k** space have been found to obey relations similar to those satisfied by E_g itself. These relations are

$$E_2{}^2 = E_{2,h}^2 + C^2, \tag{7.21}$$

$$(E_1')^2 = (E_{1,h}')^2 + (6C/5)^2, \tag{7.22}$$

$$(E_0')^2 = (E_{0,h}')^2 + C^2, \tag{7.23}$$

where the transitions are shown in the energy band schemes of the latter part of Chapter 5 and also for Ge in Fig. 7.3 of this chapter. Here $E_{2,h}$, $E_{1,h}^1$, and $E_{0,h}^1$ are the values of the interband energy in the elemental semiconductors, and these energies are regarded as a function of bond length d only. The function that is used is again a power law, but the power index depends on the interband transition in question.

The similarity between the relations (7.21)–(7.23) and the basic relation (2.23) for E_g is, of course, not accidental. The validity of (2.23), as well as (7.21)–(7.23) is most easily displayed graphically for isoelectronic sequences such as Ge–GaAs–ZnSe or diamond–BN–BeO in which the bond length d does not change significantly within the sequence. Then, since $C(\text{ZnSe}) = 2C(\text{GaAs})$, for example, $E_2{}^2$ should be linear in $(\Delta Z)^2$, where ΔZ for the $A^N B^{8-N}$ compound is simply given by $\Delta Z = 8 - 2N$. Graphs of the relations (7.21)–(7.23) are shown for the Ge-GaAs-ZnSe sequence in Fig. 7.12. From (2.20) and (2.23), similar behavior should be obtained for

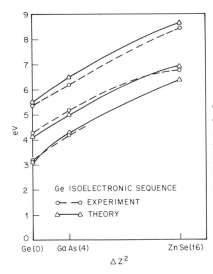

Fig. 7.12. Trends in the interband energies described by Eqs. (7.21)–(7.23) for the isoelectronic sequences Ge–GaAs–ZnSe plotted against $\Delta Z = 8 - 2N$, where N is the cation valence. In order of increasing energy the interband energies shown are E_0', E_2, and E_1'. Theoretical fit from J. A. Van Vechten, *Phys. Rev.* **187**, 1007 (1969).

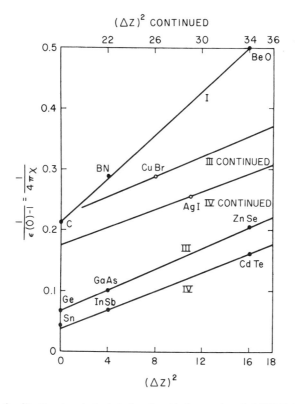

Fig. 7.13. Qualitative trends in interband optical energies of $A^N B^{8-N}$ tetrahedrally coordinated compounds are all described by the ionic energy gap C. For horizontal isoelectronic sequences (e.g., row III, Ge, GaAs, ZnSe, CuBr) C is simply proportional to $\Delta Z = 8 - 2N$. The reciprocal overall polarizability $[\epsilon(0) - 1]^{-1}$ of the crystal is proportional to $(\Delta E)^2 = (\Delta E_h)^2 + C^2$, so that $[\epsilon(0) - 1]^{-1}$ is linear in $(\Delta Z)^2$ for each sequence [from J. A. Van Vechten, *Phys. Rev.* **182**, 891 (1969)].

E_g by plotting $[\epsilon(0) - 1]^{-1}$ against $(\Delta Z)^2$. This is seen to be the case in Fig. 7.13 for isoelectronic sequences from the C, Si, Ge, and Sn rows. In gray Sn, which is semimetallic, $\epsilon(0)$ is identified with interband transitions (non-Drude contributions not associated with free carriers). The interband contribution to $\epsilon(\omega)$ is not easily separated from the Drude one as $\omega \to 0$, so Fig. 7.13 shows how to *infer* $\epsilon(0)$ in gray Sn from the known values in CdTe and InSb.

Most optical experiments measure only energy differences between electronic levels in the crystal. However, by studying the energy distribution of electrons emitted by the crystal under exposure to visible and ultraviolet light, one can also establish the position of the top of the valence band in the volume of the crystal relative to the zero of energy in vacuum well outside the crystal. This energy difference is called the ionization energy I, while the position of the bottom of the conduction band relative to vacuum is called the electron affinity A. Note that I is not defined relative to the top of the valence band near the surface, because this energy may be affected by the presence of charged impurities at or near the surface. This surface effect, which is often called band bending, will be treated in a later chapter.

The ionization energy I also is found to follow the relation

$$I^2 = I_h^2 + C^2, \tag{7.24}$$

where $I_h \propto d^{-1.3}$. On the other hand, the electron affinity A does not seem to follow a simple relation like (7.24), which suggests that I is a more basic quantity than A. From a viewpoint of adding or subtracting charge, I and A are equally significant, but the minimization of energy which determines the equilibrium lattice constant of the crystal is more strongly affected by the positions of the occupied one-electron energy levels than by the positions of the unoccupied ones, so that I should be more basic than A. Further, note that by definition

$$A - I = \Delta E_{cv}, \tag{7.25}$$

so that if I but not A follows simple chemical trends, ΔE_{cv} is likely to be irregular as well. For this reason the indirect energy gaps do not follow the simple patterns that the direct ones do.

The energy gaps E_0 and E_1 connect valence-band states with s-like antibonding conduction-band states which penetrate the atomic cores. Formulas which have been fairly successful in describing this effect are

$$E_0 = [E_{0,h} - (D - 1)\Delta E_0][1 + (C/E_{0,h})^2]^{1/2}, \tag{7.26}$$

$$E_1 = [E_{1,h} - (D - 1)\Delta E_1][1 + (C/E_{1,h})^2]^{1/2}. \tag{7.27}$$

For Ge, the parameters are $D - 1 = 0.25$, $\Delta E_0 = 10$ eV, and $\Delta E_1 = 4$ eV, which enables one to see the importance of the d-core correction terms in (7.26) and (7.27).

The interband energies which follow chemical trends, and are described by relations such as (7.21), all refer to direct energy gaps, that is, to differences between energy levels with the same value of **k**. Numerous attempts have been made to find similar formulas for differences between energy levels with greatly different values of **k**, but none have been completely successful to the kind of accuracy desired (of order 0.5 eV or less). This problem is not confined to the bond method. Empirically adjusted pseudopotentials based on optical spectra, such as those discussed in the two preceding chapters, have encountered entirely similar problems. So far the same has been true of all attempts to calculate the energy bands using more elaborate (but not necessarily more accurate) methods based on one-electron atomic or crystal potentials. The persistence and universality of this problem suggests that there must be some basic reason why the direct energy gaps follow simple chemical trends, but the indirect ones (such as that between the highest valence state at Γ and the lowest conduction states at X or L) do not.

We can obtain some insight into this puzzle from consideration of an apparently unrelated problem, that of the lattice vibration spectrum. We saw in Chapter 3 that the elastic constants, which determine the frequencies of long wavelength ($\mathbf{k} \to 0$) acoustic phonons, can be described in terms of a valence-bond model of interatomic forces. There are two important force constants, α and β, describing bond-stretching and bond-bending forces, respectively. The ratio β/α, which is a dimensionless measure of tetrahedral stability, was found to be proportional to $1 - f_i$. Thus the interatomic forces exhibit simple chemical trends as $\mathbf{k} \to 0$. However, if the same forces are extrapolated to predict acoustic frequencies $\omega(\mathbf{k})$ for large values of $\mathbf{k}$ near the faces of the Brillouin zone, the results are not accurate enough to predict trends. This is also true of optic modes.

When a phonon propagates through a crystal, it causes the ion cores to move in such a way that the valence electrons are polarized by being virtually excited to conduction band states. This polarization can be described by a dielectric polarization matrix similar to the dielectric function used to describe optical excitations. Direct optical excitations correspond to acoustic phonons in the limit $\mathbf{k} \to 0$, whereas indirect optical excitations explicitly involve phonons with large values of $\mathbf{k}$. Therefore, the facts that both direct optical excitation energies and sound frequencies are describable in chemical terms, while indirect optical excitation energies and short wavelength phonon frequencies are not describable in chemical terms, appear to be mutually consistent. Mechanical models of lattice vibration spectra suggest that forces of range longer than those involved in bond-stretching and bond-bending are required to describe $\omega(\mathbf{k})$ for large $\mathbf{k}$. These long-range forces may not involve the nearest-neighbor bonds in a simple way, and so are not expected to exhibit chemical trends.

The correlation between long-range lattice forces and one-electron energy bands which has just been discussed may puzzle the reader. The energy bands are determined from solutions of the Schrödinger equation in a unit cell of the crystal, so how can long-range lattice forces matter? The answer is that the Schrödinger equation is solved in the unit cell *subject to certain boundary conditions*. These boundary conditions are the ones imposed by translational invariance. This is equivalent to the Bloch condition $\Psi_\mathbf{k} = \exp(i\mathbf{k} \cdot \mathbf{r}) \, \mu_\mathbf{k}$, and it obviously varies considerably with $\mathbf{k}$.

We saw in the last chapter that the building up of bond charge between atoms is an interference effect which depends very sensitively on the apparent "size" of the pseudoatom core. The interference between crystal waves is different for different values of **k**, and the way that this varies with **k** is (according to the uncertainty principle) a measure of the range of the interactions involved. Thus one is not surprised to find correlations between the long range of lattice forces and the absence of chemical trends in indirect interband energies.

SPIN–ORBIT SPLITTINGS

Orbitally degenerate energy levels in crystals are split by the spin–orbit interaction $\lambda \mathbf{L} \cdot \mathbf{S}$ just as are orbitally degenerate levels in atoms. In polyvalent atoms the problem is complicated because more than one electron may contribute to $\mathbf{L} = \sum_i \mathbf{l}_i$ and $\mathbf{S} = \sum_i \mathbf{d}_i$. In the crystal, we make the one-electron approximation. This makes the calculation of the splittings straightforward once the parameter λ is known.

The magnitudes of the spin–orbit parameter λ are known for many crystals for the triply degenerate level $\Gamma_{25'}$, which splits into $J = \frac{3}{2}$ and $J = \frac{1}{2}$ levels, and for Λ, which splits into two levels. If atomic orbital language is used, as shown for the levels at Γ in Fig. 5.10, a simple relation can be obtained between the splitting Δ_0 of $\Gamma_{25'}$ and the splitting Δ_1 of Λ_3. This relation rests on the assumption that each wave function is a superposition of only p atomic orbitals, and that the p atomic orbitals to be used in wave functions at Γ are the same as those to be used in wave functions along Λ (although they are different in energy). This relation is $\Delta_1 = 2\Delta_0/3$ (the so-called "two-thirds rule"), and it is approximately satisfied in almost all the crystals for which data are available.

The experimental values of Δ_0 can be obtained in either of two ways. In one method the splitting is measured directly by adding impurities to the crystal which produce free holes at or near the top of the valence band at $\Gamma_{25'}^{3/2}$. Absorption of an infrared photon causes transitions from near $\Gamma_{25'}^{1/2}$ to near $\Gamma_{25'}^{3/2}$, so that the peak in the infrared absorption falls near Δ_0. A second method utilizes the spin–orbit splitting of $E_0 = E(\Gamma_{2'}) - E(\Gamma_{25'})$ as measured using an appropriate derivative technique such as electro-reflectance (see Fig. 7.7). This same measurement also yields the spin–orbit splitting of the E_1 peak, which gives Δ_1 (also seen in Fig. 7.7).

Theoretical values of Δ_0 and Δ_1 can be estimated as follows. Each atom A or B is assigned the spin–orbit parameter Δ_A or Δ_B, respectively. Then $\Delta_1(AB) = 2\Delta_0(AB)/3$ and

$$\Delta_0(AB) = 3\lambda/2 = (1 - f_i)\Delta_A/2 + (1 + f_i)\Delta_B/2, \qquad (7.28)$$

where one assumes that A is the cation and B the anion, i.e., $Z_A/r_A < Z_B/r_B$, as this determines the sign of $C(AB)$, according to Eq. (2.24). To use Eq. (7.28), however, one must determine the parameters Δ_A and Δ_B. These parameters can be estimated approximately [3] from atomic spectra, but there one finds that the effective value of λ (which determines Δ) varies substantially from one multiplet to another. (This variation is usually described as the consequence of configuration interaction, or the breakdown of the one-electron approximation). This means that the values of Δ_A and Δ_B in the crystal should be treated as adjustable parameters. This has been done for the values of Δ_A and Δ_B given in Table 7.2. Comparison with experiment is made in Table 7.3. The agreement is satisfactory, but not spectacular, with 12 parameters being used to fit values of Δ_0 and Δ_1 in 17 III–V and II–VI compounds. The parameters are not all independent; in each row of the periodic table they increase from left to right in much the same way.

CRYSTAL FIELD SPLITTINGS

In crystals of the wurtzite type (hexagonal symmetry) or chalcopyrite type (tetragonal symmetry) the z axis is distinguished from the x and y

TABLE 7.2

Contributions Δ_A of Atom A to the Spin–Orbit Splitting Δ_0 (eV)[a]

	Be	B	C	N	O	F
	0.002	0.004	0.006	0.009	0.010	0.010
	Mg	Al	Si	P	S	Cl
	0.01	0.024	0.044	0.08	0.09	0.09
Cu	Zn	Ga	Ge	As	Se	Br
−1.8	0.10	0.18	0.29	0.43	0.48	0.49
Ag	Cd	In	Sn	Sb	Te	I
−1.6	0.10	0.36	0.80	1.05	1.10	1.11
	Hg	Tl	Pb			
	0.5	0.9	2.0			

[a] The apparently anomalous values for Cu and Ag have to do with the admixtures of the d states of these elements into the p-like states at the top of the valence band by the crystal field. The contribution of the d states is negative, and because their density is different from that of the p-states, the effective parameters of Cu and Ag are not related in any simple way to either the p or d spin–orbit splittings of the respective atoms. The latter are discussed in Ref. 7.3. Note that the contribution of the d states is greater for Cu than for Ag because the d electrons in Cu are less tightly bound than in Ag (by 2 eV compared to 4 eV below the Fermi energy in the respective metals).

TABLE 7.3

Experimental Values of Spin–Orbit Splittings[a]

Compound	Δ_0 (eV)	Δ_1 (eV)	Δ_0 (calc)
C	0.006		0.006
Si	0.044		0.044
Ge	0.29	0.19	0.29
α-Sn		0.48	0.80
AlN			0.012
AlP			0.060
AlAs			0.29
AlSb	0.75	0.40	0.80
GaN	0.011		0.095
GaP	0.127	0.1	0.11
GaAs	0.34	0.23	0.34
GaSb	0.80	0.46	0.98
InN			0.08
InP	0.11		0.16
InAs	0.38	0.28	0.40
InSb	0.82	0.50	0.80
ZnO	-0.005		0.03
ZnS	0.07		0.09
ZnSe	0.43		0.42
ZnTe	0.93	0.57	0.86
CdS	0.066		0.09
CdSe			0.42
CdTe	0.92	0.59	0.94
HgS			0.13
HgSe		0.30	0.48
HgTe		0.66	0.99
CuF			
CuCl	-0.05		-0.15
CuBr	0.197		0.20
CuI	0.633		0.66
AgI	0.837		0.84

[a] The theoretical values are calculated from equation (7.28), using the parameters given in Table 7.2 and Table 2.2. The negative contribution of the d electrons, which was reflected in the parameters Δ_{Cu} and Δ_{Ag} in Table 7.2, is apparent in this table also to a considerably lesser extent for ZnO, ZnS, CdS, and GaN. Experimental values from M. Cardona, "Modulation Spectroscopy," *Solid State Phys. Suppl.* **11** (1969).

axes. In the absence of spin–orbit interactions the states at $\mathbf{k} = 0$ of p_x, p_y, and p_z symmetry (labeled as Γ^Y_{15} in Fig. 5.10b), are degenerate in crystals with cubic symmetry. This degeneracy is lifted by the uniaxial interactions. It is also lifted by the spin–orbit interactions. The former splits the p_z state from the p_x and p_y states, while the latter splits the $J = \frac{3}{2}$ and $J = \frac{1}{2}$ states. Both interactions taken together give rise to three energy levels. The crystal field interaction further splits the $J = \frac{3}{2}$, $m_J = \frac{1}{2}$ and $J = \frac{1}{2}$, $m_J = \frac{1}{2}$ levels. In terms of the spin–orbit parameter Δ_0 and the crystal-field parameter δ_{cryst} the three energy levels are given by

$$E_{3/2,3/2} = +\Delta_0/3,$$

$$E_+ = +\Delta_0/3 + A + (A^2 - B)^{1/2}, \qquad (7.29)$$

$$E_- = +\Delta_0/3 + A - (A^2 - B)^{1/2},$$

with $2A = -\Delta_0 + \delta$ and $B = 2\Delta_0\delta/3$.

Values of Δ_0 were given in Table 7.3. Values of δ_{cryst} inferred from Eqs. (7.29) are listed in Table 7.4 for several uniaxial crystals.

TABLE 7.4

Crystal Field Splittings in Uniaxial,
Quasi-Tetrahedrally Coordinated
Compounds

Crystal	Symmetry	δ (10^{-3} eV)
CdSe	Hexagonal[a]	41
CdS	Hexagonal	27
ZnSe	Hexagonal	61
ZnO	Hexagonal	41
ZnS	Hexagonal	55
GaN	Hexagonal	22
CdSnP$_2$	Tetragonal[b]	-100
ZnSiAs$_2$	Tetragonal	-130
CdSiAs$_2$	Tetragonal	-240

[a] Hexagonal values from B. Segall and D. T. F. Marple, "Physics and Chemistry of II–VI Semiconductors" (M. Auer and J. S. Prener, eds.), North-Holland Publ., Amsterdam, 1967.
[b] Tetragonal values from J. L. Shay, *Phys. Rev.* **B3**, 2004 (1971).

NONLINEAR SUSCEPTIBILITIES

In a strong electric field $\mathbf{F}$ (e.g., a laser beam) the dielectric susceptibility χ of a crystal is found to depend on F as follows:

$$\chi_{ij} = \chi_{ij}^{(1)} + \chi_{ijk}^{(2)}F_k + \cdots. \tag{7.30}$$

Because the nonlinear susceptibility $\chi_{ijk}^{(2)}$ is a third-rank tensor, in general it has many independent components, but in zincblende (sphalerite) crystals, there is only one independent component, just as was discussed for piezoelectric effects in Eq. (3.29). In diamond-type crystals even this component vanishes because of inversion symmetry. Thus in $A^N B^{8-N}$ zincblende crystals $\chi_{ijk}^{(2)}$ measures the ionicity f_i of the A–B bond, and the extent to which there is charge transfer between atoms A and B.

A simple model [4] shows how the charge-transfer effect can be treated in terms of the ionic part C of the energy gap, where C is given by Eq. (7.20). In zincblende crystals suppose $\mathbf{F}$ is oriented along a (111) bond; then an amount of charge ΔZ will be transferred from B to A, so that Z_B becomes $Z_B - \Delta Z$ and Z_A becomes $Z_A + \Delta Z$. Then C becomes $C + \Delta C$, where ΔC is calculated from Eq. (7.20) in terms of ΔZ, and ΔZ is related to F through the linear susceptibility $\chi_{ij}^{(1)}$. This permits the calculation of $\chi_{ijk}^{(2)}$ with no adjustable parameters. The results, which are shown in Table 7.5, are in good agreement with experiment.

One can simplify $\chi_{ij}^{(1)}$ and $\chi_{ijk}^{(2)}$ to exhibit their dependence on f_i. When this is done, the expression for $\chi_{ijk}^{(2)}$ takes the form

$$\chi_{ijk}^{(2)} \propto [f_i(1 - f_i)^5]^{1/2}, \tag{7.31}$$

TABLE 7.5

Comparison of Calculated and Experimental Values of Nonlinear Susceptibility $\chi^{(2)}$ in 10^{-7} esu for some $A^N B^{8-N}$ Compounds.

Compound	$\chi^{(2)}$(calc)	$\chi^{(2)}$(exp)
InSb	40	33
InAs	24	20
GaSb	36	30
GaAs	10	9
GaP	6.2	5.2
CdTe	7.1	8.0
ZnTe	7.3	7.3
ZnSe	3.0	2.2
ZnS	1.7	1.7

from which one can show that, all other things being equal, the maximum value of $\chi_{ijk}^{(2)}$ is attained at $f_i = \frac{1}{6}$. There are no easily obtainable zincblende crystals with f_i near this value (except for SiC, which does not readily form in the cubic structure). However, the B–C bonds in $A^{II}B^{IV}C_2^V$ compounds have values of f_i near $\frac{1}{6}$, and for this reason they are regarded as promising nonlinear optic materials.

SUMMARY

The optical spectra of semiconductors can be understood in terms of one-electron excitations. Most of these excitations conserve the crystal momentum $\hbar\mathbf{k}$, and so provide calipers for measuring vertical energy differences $E_c(\mathbf{k})-E_v(\mathbf{k})$ between conduction-band and valence-band states when $E_n(\mathbf{k})$ is plotted in the Brillouin zone. Edges and peaks in observed spectrum can be identified by combining experimental and theoretical evidence. In this way the positions of many energy levels in the highest valence band and two lowest conduction bands of many tetrahedrally coordinated $A^N B^{8-N}$ semiconductors have been determined. Variations in these energies from one crystal to another can be explained in chemical terms of covalent and ionic effects. The spectroscopic studies, on the other hand, provide a solid foundation for the chemical concepts.

REFERENCES

1. J. C. Phillips, The Fundamental Optical Spectra of Solids, *Solid State Phys.* 18 (1966).
2. D. L. Greenaway and G. Harbeke, "Optical Properties and Band Structure of Semiconductors." Pergamon, Oxford, 1968.
3. M. Cardona, "Modulation Spectroscopy," *Solid State Phys. Suppl. 11* (1969).
4. C. L. Tang and C. Flytzanis, *Phys. Rev.* **4B**, 2520 (1971).

8

Thermochemistry of Semiconductors

In discussions of bonds and bands of tetrahedrally coordinated semiconductors given in earlier chapters, emphasis has been laid on the family resemblances of all the compounds discussed. Mathematically this has meant that simple algebraic relations have been given for bond and band energies that apply to all of the $A^N B^{8-N}$ crystals. Of course, this mathematical simplicity is no accident, and there are profound structural reasons for these similarities in constitutive properties. The similarities are manifested very clearly in thermochemical properties of compounds and alloys of compounds.

There are two basic energies of a crystal that are thermochemically important. One of these is the cohesive energy required to sublimate the crystal into free atoms at standard temperature (about 300°K) and pressure (one atmosphere). Although this energy is well-defined in principle, it is known less accurately than the heats of formation in the standard state. This is represented for an $A^N B^{8-N}$ compound by $\Delta H_f^\circ(AB)$, which is zero by definition at STP when A = B. The cohesive energy $G^\circ(AB)$, on the other hand, is given by

$$G°(AB) = \Delta G°(AB) + \Delta G°(A) + \Delta G°(B), \qquad (8.1)$$

where $\Delta G°(A)$ is the cohesive energy at STP of A, and $\Delta G°(AB)$ is the Gibbs free energy of formation of the compound.

Most microscopic or quantum-mechanical discussions focus on elemental cohesive energies such as $\Delta G°(A)$. However, from a practical point of view it is $\Delta H_f°(AB)$ which is of greatest importance, since unless $\Delta H_f°(AB) < 0$, the compound will not be stable. Offhand, one might not expect to find trends in $\Delta H_f°(AB)$, because it is small in magnitude compared to $\Delta G°(A) + \Delta G°(B)$. Moreover, in order to form an A–B bond, one must first dissolve A–A bonds and B–B bonds, which may be of a quite different nature than A–B bonds. Consider, for example, ZnO. Here Zn itself in the standard state is a close-packed metal, while oxygen forms diatomic molecules. Neither structure resembles a covalent, tetrahedrally-coordinated semiconductor. Nevertheless, as we shall see, there exist well-defined chemical trends not only in $G°(AB)$, but in $\Delta G°(AB)$ or $\Delta H_f°(AB)$ as well. At present no one has succeeded in giving a general derivation of these trends, but microscopic descriptions can be made in terms of bond and band parameters which are already familiar to us from earlier chapters. Perhaps these microscopic descriptions are more useful in practical terms than a full derivation; in any case, the descriptions help us to understand thermo-chemical behavior of alloys and ternary tetrahedrally-coordinated compounds.

COHESIVE ENERGIES

In covalent crystals such as diamond, Si, Ge and gray Sn, the value of $\Delta G°(A)$ is just the cohesive energy arising from the formation of covalent bonds between the atoms. In order to be consistent with (8.1), we quote $G°(AA)$ as $2\Delta G°(A)$, in units of kcal/mole, one mole containing $2N_0$ atoms, where N_0 is Avogadro's number. For comparison with energy bands, 1 eV/atom-pair corresponds to 23 kcal/mole. There are four bonds per atom-pair. Thus in diamond, where $G° = 320$ kcal/mole, the energy per C–C bond is 80 kcal/mole.

All interaction energies become larger with decreasing bond lengths. The important question for cohesion is how rapidly this takes place. In Table 8.1 the cohesive energies for diamond-type crystals are listed together with the Fermi width $E_F = \hbar^2 k_F^2/2m$ of a valence-electron gas with density equal to that of the valence electrons in the crystal. Also shown in the table is the cohesive energy per valence electron, $G°/8$, relative to the Fermi width E_F.

TABLE 8.1

Cohesive Energies of Diamond-Type
Crystals Compared to the Fermi Energy
(or Fermi Width) of a Free Electron Gas
with Density Equal to That of the Valence
Electrons in the Appropriate Crystal

Crystal	E_F	ΔG_S	$\Delta G_S/E_F$
C	667	320	0.478
Si	287	197	0.685
Ge	265	161	0.607
Sn	210	128	0.608

From the results shown in Table 8.1 we can see that G°/E_F is approximately constant, compared to E_F, which varies by a factor of more than three between diamond and gray Sn. The cohesive energy of diamond is somewhat small, because only 2s and 2p valence states hybridize well to form directed orbitals. (It costs extra energy to promote a 2p electron to a 3d state). In Si just the reverse situation prevails: it is easy to promote a 3p electron to a 3d state, and this facilitates further hybridization and gives rise to a large, well-localized bond charge. In Ge and gray Sn the presence of d core states reduces the contribution of p–d hybridization to the bond charge. (We saw how this tendency towards metallization actually takes place in the charge density in Chapter 6). As a result, the ratio G°/E_F has about the same value in Ge and gray Sn, and this is easily understood, as the outermost core shells of s, p, and d states are occupied in both cases.

How would we expect the cohesive energy to vary in an isoelectronic sequence such as Ge–GaAs–ZnSe, or diamond–BN–BeO? Firstly, the lattice constant itself stays constant in such a sequence. This means that kinematical effects, such as those which make G° approximately proportional to E_F in diamond-type crystals, make no contribution. Thus it is the potential energy associated with $V_p(A)$ and $V_p(B)$ that is mainly important. Because $r_A + r_B$ is constant, the most important effects are connected with Z_A and Z_B. Cohesion is a binary effect and will depend on something like $(Z_A + Z_B)^2 - Z_A^2 - Z_B^2$, the latter two terms representing the free atom energies. Thus cohesive energies should be about three fourths as great in ZnSe as in Ge, and qualitatively speaking we expect the cohesive energies to decrease with increasing ionicity.

This qualitative argument is borne out by the experimental data, which are exhibited in Fig. 8.1. For each isoelectronic sequence ($r_A + r_B =$

constant), $G°$ is shown as a function of f_i. In each case the experimental data can be represented by

$$G° = G_0(\text{covalent}) - G_1 f_i, \tag{8.2}$$

where G_1 is a positive constant. In other words $G°(AB)$ correlates *quantitatively* with f_i, inasmuch as (8.2) is satisfied within the limits of uncertainties in G_0 and f_i (about 1% in each quantity).

In Chapter 2 we discussed the linear relation (8.2) with a slightly different notation [see Eq. (2.33)]. The discussion here is more complete than that given in Chapter 2, because we are able to draw on our experience of charge densities in Chapter 6. In Chapter 2 we noticed that G_1 is proportional to $d^{-1/2}$, and explained that result in terms of Thomas–Fermi screening.

PAULING'S DESCRIPTION

We saw in Chapter 2 that Pauling uses heats of formation to define electronegativity and ionicity. His approach is summarized in his famous

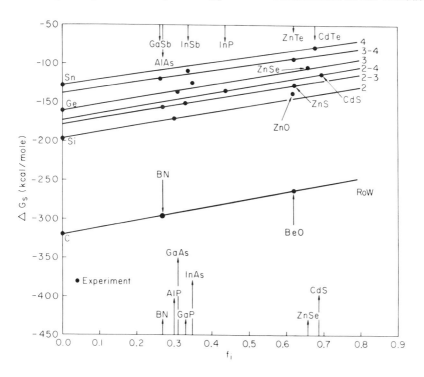

Fig. 8.1. The cohesive energies [see Eq. (8.1)] of isoelectronic sequences (bond length d nearly constant) are linear functions of the spectroscopic ionicity f_i.

equation [1]

$$-\Delta H_f^{\circ} = N_R E_I (X_A - X_B)^2 - 55.4 n_N - 26.0 n_O. \qquad (8.3)$$

Here the heat of formation of A–B bonds

$$\Delta H_f^{\circ} = D(A-B) - [D(A-A) + D(B-B)]/2 \qquad (8.4)$$

means the difference in heat of atomization of AB at STP compared to the average of the heats of atomization of AA and BB. In (8.3), N_R is the number of resonating bonds per atom-pair, about which more later; E_I is a constant, equal to 1 eV per atom-pair; X_A and X_B are empirically adjusted constants which represent elemental electronegativities (see Table 2.5); and n_N and n_O are one or zero depending on whether or not either A or B is nitrogen or oxygen, respectively.

Pauling's Eq. (8.3) has been the subject of much controversy. If it is taken literally, it is easy to fault in almost any family of crystals or molecules. Comparison with experimental data then shows large disagreements, and in some cases qualitatively wrong trends. Yet there is much that is of value in (8.3), and it actually forms an excellent point of departure for understanding the heats of formation of semiconductors. What is needed to appreciate the content of Eq. (8.3) is an analysis of the factors included in it. This analysis can be made much more precise today than was possible at the time Eq. (8.3) was introduced, because in the period of more than 30 years that have elapsed since (8.3) was devised, much more has become known about covalent systems (especially semiconductors, our special interest). The additional data confirm Pauling's general intuition, but at the same time point the way towards the refinements required to obtain satisfactory quantitative agreement with experiment.

First we must ask the meaning of the corrections for nitrogen and oxygen contained in (8.3). These are necessary, according to Pauling, to make allowance for the unusual stability of multiple bonds of these elements in their standard states (diatomic molecules). Thus N_2 is described as triply bonded, and O_2 is doubly bonded. In this way one is able to explain why the molecules NCl_3, OF_2, and Cl_2O are unstable, in spite of the stability predicted by Eq. (8.3) in the absence of the last two terms.

Pauling regards N_2 and O_2 as unusually stable because he regards all single bonds as normal. Thus in its standard crystalline state P is threefold coordinated, corresponding to three single P–P bonds about each atom. Similarly, S is twofold coordinated, corresponding to two single S–S bonds about each atom. Thus the fact that the standard states of N and O are diatomic molecules is anomalous, and some explanation is needed for the instability of NCl_3, OF_2, and Cl_2O. However, there is an alternative explanation, which has gained favor among molecular theorists [2]. This is that N_2 and O_2 are normally stable, but that single N, O and F bonds (as in the molecules mentioned by Pauling) are anomalously weak. Conventionally this is explained by "lone-pair" repulsions in these small atoms. Further discussion of the situation in molecules would lead us too far afield. However, it is worth adding that analysis of the heats of formation of crystalline transition metal carbides, nitrides and oxides again shows that N_2 and O_2 multiple bonds are normally stable, so that the last two terms in Eq. (8.3) should be discarded [3]. In molecules they must be replaced by corrections for the weakness of N and O single bonds. In crystals no corrections at all are necessary, and N and O behave normally as regards their contributions to heats of formation.

The next factor in (8.3) that warrants discussion is E_I, the universal extraionic bond energy. We know from a variety of sources that s and p valence electrons contribute similarly to hybridized bonds. This is primarily because they have similar energies and wavefunctions if the principal quantum number is the same. All the energies come about from attractive interactions of the negative electrons with positive-ion cores and repulsive interactions of the negative electrons among themselves. For example, around 1930 J. C. Slater studied atomic term values and found that these could be described by the hydrogenic formula (in rydbergs)

$$E_{nl} = - (Z^*)^2/n^2, \qquad (8.5)$$

provided that the effective nuclear charge Z^* was evaluated as follows.†
Let Z_{core} be the net positive charge (in units of e) of the nucleus plus all the core electrons with principal quantum number less than n. Let Z_s represent the screening effect of the interaction of the nl electron with other electrons of the same n. Then Z^* is given by

$$Z^* = Z_{core} - Z_s. \qquad (8.6)$$

The rules for computing Z_s treat ns and np valence electrons on the same footing, independent of n, except that 1s electrons interact with each other more strongly. The nd valence electrons, on the other hand, are found to lie "outside" the ns and np ones, or more accurately, between ns, np and $(n + 1)$s, $(n + 1)$p.

Slater's rules show that E_I should not be regarded as a constant, independent of whether one is dealing with (s, p) bonds or (s, p, d) bonds. Sometimes this difference is described as the orbital dependence of electronegativity. However, as we shall see, it is just the elemental electronegativities X_A and X_B which are the most meaningful part of Pauling's description. Therefore it is more convenient to say that it is E_I, the extra-

† For a recent application of Slater's rules, see Snyder [4].

ionic energy unit, that is orbitally dependent, and it is this energy that is expected to vary with hybridization configuration, i.e., with different atomic coordinations in different crystal structures. This is especially true when A and B represent nontransition metals and R and S are transition metals (partially filled d valence shell, or just filled d shells, as in Cu, Ag and Au). Then different constants E_{I} are required to describe A–B, A–R, and R–S bonds. Fortunately, most of the compound semiconductors with which we are concerned here contain only nontransition elements A and B. The only exceptions are semiconductors containing Cu and Ag. These indeed require special treatment at the appropriate place.

Now consider the number N_{R} in Eq. (8.3). According to Pauling, this is the number of resonating bonds present in the crystal. In some cases the rules for computing N_{R} become complicated, but in tetrahedrally coordinated crystals $A^N B^{8-N}$ (for which $A \neq$ Cu or Ag), one has $N_{\mathrm{R}} = N \leq 4$. In an isoelectronic sequence such as Ge–GaAs–ZnSe Pauling's electronegativities (Table 2.5) make $X_{\mathrm{A}}-X_{\mathrm{B}} \propto 8-2N$. Thus the extra-ionic term in (8.3) is proportional to

$$\Delta H_{\mathrm{f}}{}^{\circ}(\text{isoelectronic}) \propto N(4 - N)^2. \qquad (8.7)$$

To make the bond more ionic, we decrease N from 4(Ge) to 3(GaAs) to 2(ZnSe). Let $M = 4-N$, then (8.7) can be written

$$\Delta H_{\mathrm{f}}{}^{\circ}(\text{isoelectronic}) \propto (4-M)M^2. \qquad (8.8)$$

If it were not for the factor $4-M = N_{\mathrm{R}}$, $\Delta H_{\mathrm{f}}{}^{\circ}$ would increase quadratically with M, and $\Delta H_{\mathrm{f}}{}^{\circ}(\text{ZnSe})$ would be four times as large as $\Delta H_{\mathrm{f}}{}^{\circ}(\text{GaAs})$. The effect of the factor N_{R} is to make $\Delta H_{\mathrm{f}}{}^{\circ}$ saturate as M approaches 4. When $M = 4$, the B atoms are of the rare-gas type (Kr, in the specific case of the Ge isoelectronic sequence), and clearly in this limit the covalent bonding energy must vanish. This is the purpose of the factor N_{R} in (8.3): it describes the approach to closed-shell, or rare-gas behavior, in a way consistent with the classical theory of chemical valence.

There is another way of viewing Eq. (8.8), and this way is more appropriate to understanding the heats of formation of semiconductors. One may write simply

$$\Delta H_{\mathrm{f}}{}^{\circ} \propto M^2 - M^3/4 \qquad (8.9)$$

and then regard (8.9) as a Taylor series expansion in powers of M. The constant term of this series is zero by definition. The coefficient of the linear term has been set equal to zero because the extra-ionic energy is expected to be quadratic in M, while the coefficient of M^3 is determined by classical valence considerations ($\Delta H_{\mathrm{f}}{}^{\circ} = 0$ when $M = 4$). However, these assumptions are rather crude, and it should be possible to improve on them.

By now the reader may be thinking that my intention was to damn Eq. (8.3) with faint praise. After all, each and every factor in it has been scrutinized and found wanting in some respect, with the possible exception of the elemental electronegativities X_A and X_B, over which the cloud of orbital dependence still hovers. However, this is not really the point. In most cases $\Delta H_f{}^\circ$ is very small compared to the cohesive energy G_a. Therefore most solid state physicists have argued that it is not possible to understand $\Delta H_f{}^\circ$ at all, since quantum mechanical calculations yield G_a, and with very few exceptions the calculations are not sufficiently accurate to yield useful values for $\Delta H_f{}^\circ$. Yet when we come to study the experimental data, there are obvious patterns and trends in $\Delta H_f{}^\circ$ which the quantum theorists told us could not be understood.

The great merit of Eq. (8.3) is therefore twofold. Firstly, it focuses on the extra-ionic energy, which is most probably the largest term in $\Delta H_f{}^\circ$ in most semiconductors, and secondly, it tries to analyze the factors which determine the trends in this energy. Therefore it functions as a touchstone, as an example of the kind of equation which we shall have to construct if we wish to understand thermochemical data on semiconductors. In fact, Eq. (8.3) gives about as useful a description of $\Delta H_f{}^\circ$ as one can expect, if one insists on using one and the same equation to describe all molecules and all crystals. But there is no reason for restricting our understanding in this way and insisting on using only one equation. As Pauling states, the purpose of introducing the electronegativity table is to measure the power of an atom *in a bonded state* to attract electrons to itself. In this way he frees himself from atomistic (free atom) concepts, and avoids all the problems which arise because of the very great changes in valence-electron density which occur when isolated atoms are partially compressed together in molecules or even more compressed together in crystals. (See also the quantum-mechanical discussion of this problem given in connection with Fig. 6.3).

The next logical step in understanding $\Delta H_f{}^\circ$ is thus to look for equations similar to (8.3), but to let the form of the equation vary depending on the family of crystal structures in question. (Chemists might object that molecules do not fall neatly into well-defined, separate families, so that this descriptive approach will not work so easily in that case. True, but here we are interested in semiconductors). If this is to be our approach, the family of tetrahedrally coordinated $A^N B^{8-N}$ semiconductors should be the one that we consider first. According to Fig. 1.3, the bonding scheme here is particularly simple, with completely occupied sp^3 bonds and no electrons in non-bonding or antibonding orbitals. Moreover, there are many compounds in the family, and because of its technological importance accurate values for $\Delta H_f{}^\circ$ are available in many cases.

IONICITY AND METALLIZATION

Charge transfer gives rise to electrostatic attractions between cations and anions, and therefore it is an obvious mechanism for cohesion in both molecules and crystals. In crystals, however, it is often the case that the cations and anions arrange themselves on separate sublattices, and the resulting Coulomb interactions are larger, for example, in a diatomic NaCl crystal than they would be in a diatomic NaCl molecule. Because of the long range of the Coulomb interaction, it is possible that in cases of weakly ionic binding, it is just this long-range attraction that tips the scales in favor of compound formation, and makes $\Delta H_f°$ negative. In a crystal this long-range part of the energy is present if the structure is predominantly covalent. However, in the crystal, there is also the possibility that the compound will be a metal, in which case the long-range part of the ionic energy will be lost.

We have already discussed in Chapters 4 and 6 some of the consequences of the tendency towards metallization which is found with increasing atomic number. In elements from column IV of the periodic table, Ge ($Z_A = 32$) has the diamond structure, Pb ($Z_A = 82$) is a close-packed metal, while Sn ($Z_A = 50$) is allotropic, having either the gray form (diamond structure) or white form (metallic). The trend towards metallic behavior has an effect on the heats of formation of tetrahedrally-co-ordinated semiconductors, just as the trend towards closed-shell cations and anions (NaCl structure) is mainly responsible for the heat of formation itself.

Our problem now is to combine these two trends in a single formula, containing as few adjustable parameters as possible, that will give an accurate description of the experimental data, and at the same time enable us to understand quantitatively how ionicity and metallization affect $\Delta H_f°$.

HEATS OF FORMATION

One of the reasons Pauling chose to apply (8.3) to all partially-ionic molecules and crystals was that in this way his approximately 100 adjustable electronegativity parameters (one for each element) were over-determined. To reduce the number of adjustable parameters in our formula for $\Delta H_f°$, we take the parameters for each compound $A^N B^{8-N}$ from its known spectroscopic energy levels. At first sight, this may not seem to be a gain, because one has to have the compound (and therefore know its heat of formation, at least in principle) to determine its energy levels. However, that is not quite so. For example, in defining the average covalent and ionic energy gaps, E_h and C, respectively, we found that E_h scaled like $d^{-2.5}$ and that C was given by the simple formula, Eq. (2.24). Moreover, the individual energy gaps which are needed to describe metallization also can be predicted from algebraic relations such as Eqs. (7.21–3), (7.26) and (7.27). Finally, we can relate many of these quantities to the atomic pseudopotentials with equations like (6.21) and (6.22). In this way the heats of formation are traced back to quantum-mechanical properties, at least on an inductive or descriptive basis. Of course, no one has yet found rigorous derivations for all these relations, because no one knows how to calculate ΔH_f° quantum-mechanically, and even if one did, one might not be able to isolate in the computer the physical mechanisms described here. Perhaps, after all, it is the mechanisms that are most important to us, and so-called rigorous mathematics should be left as homework exercises for freshman calculus students.

The spectroscopic formula [5] for ΔH_f° has the form

$$-\Delta H_f^\circ = d^{-3}(\text{ionicity})\,(\text{metallization}) - 20(n_{Cu} + n_{Ag}/2). \quad (8.10)$$

We postpone for awhile the discussion of the corrections for Cu and Ag, and concentrate our attention on the first term on the right-hand side of Eq. (8.10). The bond length d enters through the scaling factor d^{-3} for the following reason. At STP the constituents A and B are either metallic or only weakly covalent, except for $N = 4$, where $\Delta H_f^\circ = 0$ by definition. Therefore ΔH_f° is connected with the ionic part of the energy gained from the energy gap E_g between bonding and antibonding states. As we saw in Fig. 5.5, a fraction of states $E_g/4E_F$ is actually involved in the bonding, and the energy gained by each of these states because of bonding is $E_g/2$. Thus ΔH_f° is proportional to E_g^2/E_F. However, E_g scales approximately like $d^{-2.5}$, while $E_F \propto d^{-2}$. This gives $E_g^2/E_F \propto d^{-3}$, and explains the first factor in (8.10).

We now turn to the ionicity factor in (8.10). This could be proportional to a term of the form

$$(\text{ionicity}) = \sum_{j=1}^{m} a_j(f_i)^j \quad (8.11)$$

where f_i is the spectroscopic ionicity,

$$f_i = C^2/(E_h^2 + C^2). \quad (8.12)$$

[Notice that (8.12) already saturates for $C \gg E_h$]. Actually, however, we have seen several instances in which properties are linear in f_i (cohesive energy, Fig. 8.1; shear strength, Fig. 3.2). Therefore we choose for the (ionicity) factor simply f_i. For an isoelectronic sequence such as Ge–

GaAs–ZnSe, E_h is practically constant, and C is proportional to $M = 4 - N$. Thus (8.12) has the form

$$f_i \propto M^2/(\text{const} + M^2) \qquad (8.13)$$

a relation quite different from the one posited by Pauling, Eq. (8.8). We mentioned after Eq. (2.31) that Pauling's approach cannot reconcile ratios of ΔH_f° ($M = 2$) to ΔH_f° ($M = 1$). The basic reason for this lies in the difference between Eqs. (8.8) and (8.13).

We now come to the (metallization) factor in Eq. (8.10). This factor becomes significantly different from unity only in compounds containing atoms from the Ge, Sn, or Pb rows. In discussing the energy bands of gray Sn and the Hg chalcogenides (Chapter 5, p. 119), we saw that metallization meant dehybridization or overlap of bonding p and antibonding s states. The charge densities of these two kinds of states were discussed in Chapter 6. The bonding p states pile up charge at the bond sites, while the antibonding s states have zero (or nearly zero) charge densities at the bond sites.

At the point in the reduced Brillouin zone labeled Γ ($\mathbf{k} = 0$) one has enough symmetry so that, at least in the absence of spin–orbit coupling, there is no mixing of antibonding s states into the bonding p states (they have group symmetry Γ_1 and Γ_{15} respectively, in sphalerite crystals; see Fig. 5.10). However, at most points in the Brillouin zone (and of course all points contribute equally to determining the valence-charge density), the smaller the average energy difference $\bar{E}$ between bonding p and antibonding s states, the greater their mixing, and the weaker the bonding.

Optimal hybridization or bonding should take place when $\bar{E}$ is approximately equal to E_g. Study of the energy bands of these crystals, as described by Eqs. (7.26) and (7.27) compared to (7.21–7.23), suggests that we define

$$\bar{E} = (E_0 + E_1)/2, \qquad (8.14)$$

where E_0 and E_1 represent orbital energies, averaged over spin–orbit split multiplets. The corresponding gaps are those described by Eqs. (7.26) and (7.27). Dehybridization is described by the term $(E_g/\bar{E})$ when $\bar{E}$ is less than E_g, and by unity otherwise. An appropriate form for the metallization factor is

$$(\text{metallization}) = 1 - b(E_g/\bar{E})^2, \qquad (8.15)$$

where the parameter b is adjusted according to

$$b = 0.05 = (\bar{E}/E_g)^2 \text{ gray Sn}, \qquad (8.16)$$

because gray Sn is allotropic with white Sn.

We are now ready to quote the formula for ΔH_f°. It is[5]

$$\Delta H_f^\circ(\text{AB}) = \Delta H_0 f_i (d_{\text{Ge}}/d_{\text{AB}})^3 [1 - b(E_g/\bar{E})^2{}_{\text{AB}}] + 20.0 n_{\text{Cu}} + 10.0 n_{\text{Ag}},$$

$$(8.17)$$

where $\Delta H_0 = -68.6$ kcal/mole. All we need explain now is why corrections are needed for Cu and Ag in Eq. (8.17).

The cohesive energies of elemental Cu and Ag exceed those of the monovalent alkali metals by about 40 and 20 kcal/mole, respectively. (The excess is less for Ag than for Cu, because the d electrons in Ag are twice as tightly bound, relative to the Fermi energy. This is the reason for the difference in color of the two metals). The way the d electrons contribute to the cohesive energies of the metals is through exchange of d electrons between atoms, via virtual states above E_F. This exchange is reduced now that the Cu or Ag atoms are only second-nearest neighbors, rather than nearest neighbors. Empirically about half this d electron energy is lost, which reduces the magnitude of $\Delta H_f^\circ < 0$, as indicated in Eq. (8.17).

Before comparing Eq. (8.17) with experiment, one should notice that (8.17) is remarkably free of adjustable parameters. The constant factor ΔH_0 has been adjusted, but b has been assigned a plausible value in Eq. (8.16). Certainly there are no adjustable parameters which vary from element to element, or compound to compound. Rather one may say that the adjustable parameters contained in Eq. (8.17) describe characteristic features of the medium.

Comparison is made in Table 8.2 between experiment, the values predicted by Pauling's Eq. (8.3), and the spectroscopic predictions of Eq. (8.17). It is seen that the values predicted spectroscopically are in remarkably good agreement with experiment (probably to within the limits of experimental errors in ΔH_f°). This shows that we have achieved our goal of identifying the mechanisms which determine ΔH_f°.

The case of the Hg chalcogenides (HgS, HgSe, and HgTe) requires separate discussion. It is in these compounds that the effects of metallization are most dramatic, both on bond length and on heats of formation. The effects are illustrated in Table 8.3, which compares these quantities in CdX with those in HgX ($X = S$, Se, and Te). Because Hg is heavier than Cd, one would expect that its covalent radius should be larger, and one would expect that the lattice constant of HgX would be greater than that of CdX. For example a/a_0 is greater in PbTe than in SnTe by 0.27. But according to Table 8.3, a/a_0 is less in HgTe than in CdTe by 0.09. Thus Hg has "contracted" relative to Cd. This is the only irregularity of this kind found in tetrahedrally coordinated $A^N B^{8-N}$ semiconductors. It is directly related to dehybridization or metallization effects arising from overlap of p bonding states with s antibonding states.

TABLE 8.2

The Heats of Formation ΔH_f° of $A^N B^{8-N}$ Tetrahedrally-Coordinated Crystals[a]

Crystal	$-\Delta H$ (Pauling)	$-\Delta H$ (spectro.)	$-\Delta H_{expt}$
BN	13.6	63.6	60.8
BeO	158.0	128.5	143.1
AlP	24.8	22.2	39.8
GaAs	11.0	16.3	17
ZnSe	29.4	39.8	39
InSb	2.8	9.0	7.3
CdTe	4.4	25.1	22.1
ZnO	140.1	72.5	83.2
AlAs	17.3	17.3	27.8
GaP	17.3	24.8	24.4
ZnS	37.3	42.1	49.2
InP	11.0	20.3	21.2
CdS	29.4	35.8	38.7
AlSb	11.0	20.3	
GaSb	6.2	9.5	10.0
ZnTe	11.5	25.9	28.1
InAs	6.2	11.5	14.0
CdSe	22.5	30.9	32.6
CuI	8.3	14.8	16.2
CuBr	18.6	22.7	25.0
CuCl	27.8	34.9	32.8

[a] Comparison is made between the experimental values, Pauling's values, and the values
given by the spectroscopic Eq. (8.17) of the
text. All energies are in kcal/mole.

TABLE 8.3

Heats of Formation and Effective Cubic Lattice
Constants in Cd and Hg Chalcogenides

X^a	$-\Delta H_f^\circ$ (kcal/mole)		$a_{eff}/a_0{}^b$	
	CdX	HgX	CdX	HgX
S	38.7	12.8	11.05	11.07
Se	32.6	11	11.49	11.51
Te	22.1	10	12.25	12.16

[a] Here X is a chalcogenide (X = S, Se, or Te).
[b] In units of the Bohr radius a_0.

To see that this is the case, it is instructive to assume that the lattice constants a/a_0 of HgX are greater than those of CdX by 0.27. With these hypothetical lattice constants one can calculate E_h, C, E_{0h}, and E_{1h}, and apply Eqs. (7.21)–(7.27) to predict the values that the quantities entering (8.17) would have in those hypothetical structures. One finds immediately that $\bar{E}$ would be much less than it is in gray Sn, which means that the metallization factor in (8.17) would reverse sign and $\Delta H_f^\circ > 0$; i.e., the compound would be unstable at the hypothetical lattice constant.

Does this mean that Hg chalcogenides should not form at all? Not necessarily. We saw in Chapter 6 that under pressure the energy gaps E_0 and E_1, which determine $\bar{E}$ through Eq. (8.14), increase rapidly. Thus the energy lost by the HgX compounds because of dehybridization can be recovered, at least in part, by contraction or by reduction of the lattice constant. Detailed analysis [5] shows that in each case E_0 is reduced to a value near that found in gray Sn (almost 0.0 eV) rather than the value close to -1.0 eV which would have obtained at the hypothetical lattice constant.

A consequence of the internal contraction of the Hg chalcogenides is a great reduction in ΔH_f° compared to ZnX and CdX compounds. Moreover, in the latter, as we go from X = Te to X = Se to X = S, at each step ΔH_f° increases by 10 kcal/mol. But in the HgX sequence the corresponding changes are only 1 kcal/mole, the remainder of the energy being used in internal contraction.

ENTROPIES OF FUSION

It is customary, in discussing the entropy of fusion ΔS^F, to separate ΔS^F into two contributions, one electronic and one vibrational. The temperature of fusion T^F is small compared to E_F/k_B, where E_F is the Fermi width of the valence electron gas and k_B is Boltzmann's constant, because E_F/k_B is of the order 10^5 °K. This means in conventional terms that most of ΔS^F is associated with the "softness" of the liquid compared to the solid, which increases the number of vibrational states present in the liquid, and hence increases its entropy relative to that of the solid.

This approach is correct as far as it goes, but it does not go far enough to explain systematic quantitative differences between ΔS_m^F for metals and ΔS_t^F for tetrahedrally coordinated semiconductors. Experimentally it is found that ΔS_t^F is two to three times greater than ΔS_m^F, and conventionally this would be explained through the greater stiffness of the tetrahedral covalent crystal compared to the metallic crystal. Of course, this still leaves the calculation of the vibrational spectrum as a lengthy task, according to the discussion in Chapter 4.

TABLE 8.4

Melting Points T^F for $A^N B^{8-N}$ Semiconductors[a]

Crystal	T^F (°K)	Crystal	T^F (°K)
C	~4400	HgSe	1072
Si	1685 ± 2	HgTe	943
Ge	1210 ± 1		
α − Sn	428 ± 24	BeTe	≲1570
β − Sn	508	MgTe	
		ZnTe	1563
SiC	2810	CdTe	1365
BN	?	BeSe	?
AlN	~3300	MgSe	?
GaN	M2000	ZnSe	1788 ± 20
InN	~1370	CdSe	1512
BP	~2300	BeS	?
AlP	?	MgS	?
GaP	1740	ZnS	2196
InP	1343	CdS	1825 ± 30
BAs	?	BeO	?
AlAs	2013 ± 20	ZnO	?
GaAs	1511		
InAs	1216	CuCl	703
		CuBr	761
AlSb	1353	CuI	861
GaSb	985	AgI	831
InSb	809		

[a] Experimental values are given here. Theoretical calculations of these values of T^F based on the dielectric theory of covalent bonding have been made by J. A. Van Vechten, *Phys. Rev.* **7B**, 1479 (1973).

A much simpler approach to this thermochemical problem, based on bonding concepts, has been suggested by Van Vechten. He separates ΔS_t^F into two parts, ΔS_m^F and ΔS_b^F, where ΔS_b^F is the bonding contribution to the entropy of fusion of the covalent crystal. As we saw in Chapter 6, the effect of bonding is to localize electrons in the bonding region between atoms, while leaving almost no valence charge in the antibonding or interstitial regions of the unit cell. With N atoms there are $4N$ valence electrons, which occupy $2N$ bonding states while leaving $2N$ antibonding states vacant. It is this separation of the bonding and antibonding states, of course, that accounts for the increased stiffness of the covalent crystal compared to the metallic one. But this same separation leads to a value for

$\Delta S_b{}^F$ of $4Nk \ln 2$, a value which is in good agreement with experiment. The value of $\Delta S_m{}^F$ is empirically found to be about 1.1 Nk, which is similar to the entropies of melting of metals such as Mg and Cu.

With values for $\Delta S_t{}^F$ and $\Delta H_t{}^F$ (which is calculated according to ideas similar to those discussed in the last section), one can calculate the melting points T^F of $A^N B^{8-N}$ semiconductors. The experimental values of T^F are listed in Table 8.4 for the reader's convenience; the theoretical values are in good agreement with the experimental ones.

THE PbS OR $A^N B^{10-N}$ FAMILY

We saw in Chapter 1 that diatomic compounds such as PbS, with ten valence electrons per atom pair, have the occupied orbital configuration $s^2 p_+{}^3$, i.e., with nonbonding s orbitals and bonding p orbitals, each doubly occupied by electrons of opposite spin. The members of this ten-electron family include GeTe, SnSe, SnTe, PbS, PbSe, and PbTe, as well as As, Sb, and Bi. The crystal structures in this family are typically those of NaCl, with six nearest neighbors. In As, Sb, Bi, and GeTe the cubic structure is slightly distorted into a rhombohedral one, with three nearer near-neighbors and three further near-ones. The distortion angles δ are shown in Fig. 8.2 plotted against the dielectric electronegativity difference $\Delta X = X_A - X_B$. Increasing values of ΔX stabilize the NaCl structure with $\delta = 0$. The elemental crystals As, Sb, and Bi are semimetals, with very small overlap between occupied and unoccupied bands, and a very small density

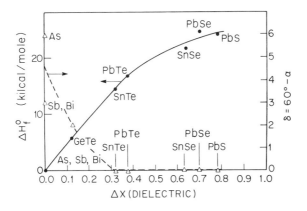

Fig. 8.2. Rhombohedral distortion angles δ and heats of formation $\Delta H_f{}^\circ$ in crystals belonging to the PbS family, plotted against the dielectric electronegativity difference, taken from Table 2.5.

of states near the Fermi energy. Some of these states are split by the spin–orbit interaction, which is much larger for atoms from the Sb and Bi row than for the As row (see Table 7.2). This is the reason for the difference in δ between Bi and Sb on the one hand and As on the other hand.

The most interesting structural question for the $A^N B^{10-N}$ family concerns the relation between the heats of formation $\Delta H_f{}^\circ$ and the electronegativity difference ΔX. This is also shown in Fig. 8.2. The sequence in GeTe, SnTe, and PbTe is especially striking: it indicates that $\Delta H_f{}^\circ$ is *linear* in $|\Delta X|$, in contrast to essentially quadratic relation found in $A^N B^{8-N}$ compounds. Thus the linear behavior of $\Delta H_f{}^\circ$ is the new feature characteristic of the $A^N B^{10-N}$ family.

The central difference between $A^N B^{8-N}$ sp^3 bonds and $A^N B^{10-N}$ p^3 bonds is that the latter are accompanied by nonbonding s electrons. These s electrons repel one another (closed s subshell), and it seems likely that this repulsion is maximum when $\Delta X = 0$. In effect, $|\Delta X| > 0$ lifts a "degeneracy" between s^2 subshells on the anion and cation, and the lifting of this degeneracy gives rise to a contribution to $\Delta H_f{}^\circ$ which is first-order in $|\Delta X|$. In $A^N B^{8-N}$ compounds, on the other hand, in the limit $\Delta X = 0$ one still has $E_g = E_h > 0$, so that contributions to $\Delta H_f{}^\circ$ are at least of order $(\Delta X)^2$.

An aside on electronegativity scales is in order here. According to Pauling's universal scale (Table 2.5), $X(\text{Ge}) = X(\text{Sn}) = X(\text{Pb}) = 1.8$, so using his scale one would not find the trends in the tellurides shown in Fig. 8.2. The dielectric scale is based on the properties of $A^N B^{8-N}$ semiconductors, and it is still useful for describing small differences in properties of the $A^N B^{10-N}$ semiconductors.

PRESSURE-INDUCED PHASE TRANSITIONS

Perhaps the most remarkable aspect of tetrahedrally-coordinated structures is their low density. It is not surprising, therefore, that under pressure tetrahedrally-coordinated semiconductors can be transformed to structures with higher density. Typical values of $\Delta V/V_t$ (where V_t is the molar volume of the tetrahedrally coordinated structure at the transition pressure P_t, and ΔV is the volume change at the transition) are about 0.2. Phase changes for metals or insulators usually involve fractional volume changes of a few percent, or about ten times less than the volume changes associated with collapse of tetrahedral structures.

The higher density structures to which tetrahedrally-coordinated $A^N B^{8-N}$ compounds transform under pressure are the ones already familiar to us from discussions of chemical trends at zero pressure. Broadly speaking,

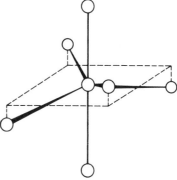

Fig. 8 3. The nearest-neighbor coordination configuration in the white Sn structure. There are four nearest neighbors, displaced from the xy plane toward tetrahedral sites, and two slightly more distant ($+5\%$) neighbors along the z axis.

there are two possible higher-density structures, one insulating and of the NaCl type, and one metallic and of the white Sn type (see Fig. 8.3). Both structures are approximately sixfold coordinated, and both are about 20% more dense than the tetrahedral structure. The transitions can be classified diagramatically by making a thermochemical plot of the kind shown in Fig. 8.4. Here the temperature has been fixed at some value well below the melting points of the crystals involved (say at room temperature). The abscissa is the spectroscopic ionicity (a chemical variable) and the ordinate is the pressure (typically of order 100 kbar at P_t), a conventional thermo-dynamic variable. Because ionicity is a well-defined quantity for $A^N B^{8-N}$

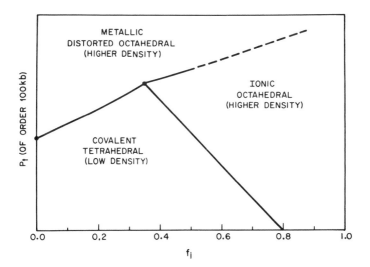

Fig. 8.4. Schematic thermochemical (P versus f_i) phase diagram showing the co-valent, ionic, and metallic structures of $A^N B^{8-N}$ compounds.

crystals, this plot can be more useful in certain situations than a conventional thermodynamic P vs T plot.

The first feature that we notice from Fig. 8.4 is that for $f_i \lesssim 0.35$, under pressure the semiconductor transforms directly to a metallic structure, while for $0.35 \lesssim f_i \leq 0.785$, the semiconductor first transforms to an insulating NaCl structure, and transforms to a metal only at much higher pressures. (The second transition has actually been observed only in a few cases, such as CdTe). These qualitative features are common to all $A^N B^{8-N}$ compounds. The only fixed point in Fig. 8.4, however, is the one corresponding to the covalent-ionic transition at $P_t = 0$, $f_i = 0.785$. We have previously examined the meaning of this point by plotting populations of $A^N B^{8-N}$ compounds in the (E_h, C) or (covalent gap, ionic gap) plane in Fig. 2.7. Now we can examine the covalent-ionic transition in more detail. We use the transition pressures P_t and the volume discontinuities $\Delta V/V_t$ to compute differences in molar free energies of compounds with ionicity f_i near the critical ionicity. This corresponds to the line separating the covalent and ionic structures in Fig. 8.4, and in particular the segment of the line near the point $P_t = 0$ in the ionicity range $0.6 \lesssim f_i \leq 0.8$.

The Gibbs free energy difference ΔG_t between a compound which has the tetrahedral structure at $P = 0$ and the NaCl structure for higher pressures is nearly given by $P_t \Delta V/V_t$ (in kcal/mole). This is because the compressibilities are nearly equal in the two structures. Before plotting this energy difference against the dimensionless variable f_i, however, we need to reduce the energy difference itself to dimensionless units. All energies are larger for smaller bond lengths, of course, while f_i is confined to the range $0 \leq f_i \leq 1$ regardless of bond length.

The natural energy to use for reduction purposes is the energy of atomization defined by Eq. (8.1). In Fig. 8.5 the dimensionless ratio $\Delta G_t(\text{AB})/G°(\text{AB})$ is plotted as a function of f_i for $0.6 \leq f_i \leq 0.8$. Quite a good fit is obtained for the data points, with the exception of CdS. The measured value of P_t (CdS) varies considerably from sample to sample, depending on the way in which the crystal was grown, so this one deviation may not have much significance.

It has been traditional, within the framework of billiard-ball models of crystal structure, to explain the covalent-ionic transition in terms of the ratio of cation to anion radii. (See "historical note" at the end of Chapter 2). In the billiard-ball model it is argued that when the anions become much larger than the cations, then anion–anion contact will occur. With an anion radius r_B and a cation radius r_A, anion–anion contact occurs in the NaCl structure for $r_B/r_A > 1 + (2^{1/2})$. Then a transition is supposed to take place to a tetrahedral structure in which the anions are no longer in contact. But regardless of one's definition of ionic radii, it is clear that

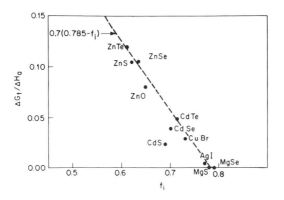

Fig. 8.5. The reduced free energy difference between $A^N B^{8-N}$ compounds in the ionicity range $0.6 \le f_i \le 0.8$. The difference in energy at $P = 0$ between the compound in the NaCl structure and in a tetrahedrally coordinated structure is measured by $P_t \, \Delta V / V_t$, as discussed in the text.

r_A / r_B must be much larger in ZnO than in ZnTe. Nevertheless, according to Fig. 8.5, $\Delta G_t / G^\circ$ scales with f_i, and it certainly would show no simple behavior as a function of r_A / r_B.

We now turn to the region $0 \lesssim f_i \le 0.35$ of Fig. 8.4. Here we have a simple covalent-metallic transition, again with $\Delta V / V_t$ approximately equal to 0.2, with the metallic structure being nearly sixfold coordinated. As one can see in Fig. 8.3, the white Sn structure still shows vestiges of tetrahedral coordination, the four atoms near the xy plane being displaced from the plane towards tetrahedral positions. These atoms are actually the four nearest neighbors, the two neighbors along the z axis being 5% further away. However, one should still think of the white Sn structure as approximately sixfold-coordinated, because its density is 20% less than that of gray Sn, just as was the case for the NaCl-structure forms of the compounds shown in Fig. 8.5.

The transition pressure for the covalent-metallic transition in the range $0 \lesssim f_i \le 0.35$ is determined by two terms, one of which is the "pure" covalent–metallic transition energy found in diamond-type crystals, and the other of which involves the reduction of ionic binding energy in the metallic state. A quantum-mechanical theory of the "pure" covalent-metallic energy ΔG_{cm} can be given, but here we simply assume that ΔG_{cm} is a smooth function of bond length d, just as the covalent energy gap E_h is [see Eq. (2.21)]. This function is known at Si, Ge and gray Sn, and a smooth curve drawn through the known points gives $\Delta G_{cm}(d)$. We then assume that a constant fraction x of the extraionic energy is lost when

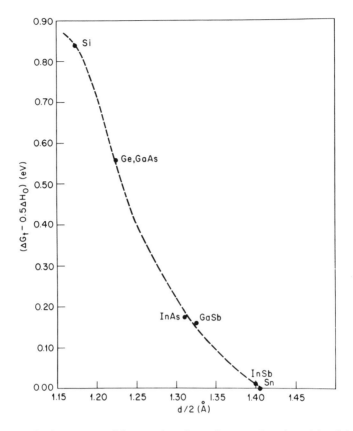

Fig. 8.6. The free energy difference function ΔG_{cm} as a function of bond length d for predominantly covalent compounds ($0 \leq f_i \leq 0.35$) which make a covalent–metallic transition under pressure (see Fig. 8.4).

the compound passes to the metallic state. (Of course, some of the ionic energy will remain because of nearest-neighbor interactions, even in the metallic state). The molar transition energy ΔG_t is then given by

$$\Delta G_t(\text{AB}) = \Delta G_{cm}(d_{\text{AB}}) + x\Delta H_f{}^\circ(\text{AB}), \qquad (8.18)$$

where the ionic energy is identified with the heat of formation of the AB compound. Analysis of the experimental data shows that the choice $x = 0.5$ gives a good fit to ΔG_t for AB = GaAs, InAs, GaSb and InSb. This is shown in Fig. 8.6. Although the compounds InAs and GaSb are isoelectronic, their ionicities and heats of formation are quite different, so the fact that Eq. (8.18) places them both at the same point on the curve for $\Delta G_{cm}(d)$, which interpolates between Ge and gray Sn, is quite pleasing.

IDEAL SOLUTIONS

The thermochemistry of many semiconductor alloys is relatively simple. Ordinarily one expects to be able to treat accurately and simply only mixtures of dilute, nonpolar gases, while dense gases and liquids cannot be so treated because they usually do not dissolve appreciably in one another. In crystals, which are denser than liquids, there are in addition the possibilities of compound formation, as well as a variety of possible crystal structures. Why then do some semiconductor compounds dissolve completely in one another; that is, why are they miscible in all proportions?

A few elemental alloys are known which exhibit simple thermochemical behavior. The two most notable examples are Cu–Ni alloys and Ge–Si alloys, whose crystallization curves are shown in Fig. 8.7 and Fig. 8.8, respectively. If the two elements have the same melting temperature T_0 and the same heat of melting ΔH_m, and the heat of mixing is zero, then A and B become thermodynamically indistinguishable, and the alloys will also melt at the same temperature T_0, regardless of composition x_A, where x_A is the molar fraction of constituent A and $x_B = 1 - x_A$ is the molar fraction of constituent B.

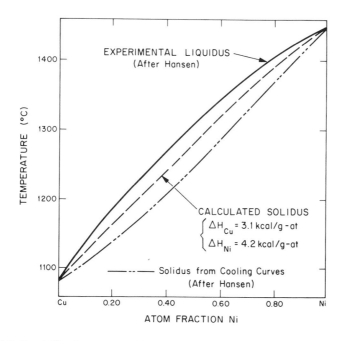

Fig. 8.7. Crystallization curves of Cu–Ni alloys. The solidus curves is calculated from ideal solution theory by means of Eq. (8.23) [6].

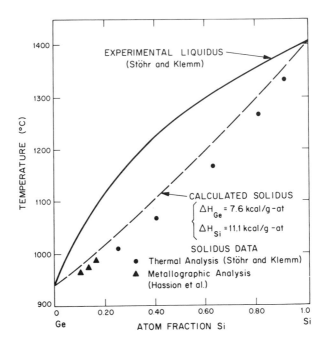

Fig. 8.8. Crystallization curves of Ge–Si alloys. Again the solidus curve is calculated from ideal solution theory by Steininger [6].

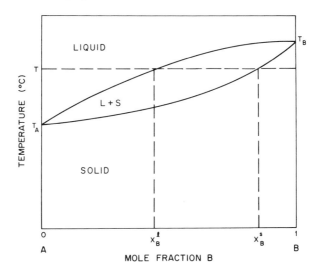

Fig. 8.9. A sketch of crystallization curves for an ideal solution [6].

The slightly more complicated behavior of Cu–Ni and Ge–Si alloys, which is shown in Figs. 8.7 and 8.8, means the following (see Fig. 8.9). At a temperature T_0 such that $T_0(A) < T_0 < T_0(B)$, a liquid alloy and a solid alloy coexist in thermodynamic equilibrium. However, their compositions $x_B{}^l$ and $x_B{}^s$ are different: the composition of the liquid alloy is given by the intersection of the horizontal line T with the liquidus curve, while the composition of the solid alloy is determined by the intersection with the solidus curve. These curves can be approximated by assuming that both the liquid alloy and the solid alloy are ideal systems, so that the total free energy G is additive:

$$G = N_A g_A + N_B g_B, \tag{8.19}$$

where g_A and g_B are the molar free energies of elements A and B, and N_A and N_B denote the number of moles of A and B which are present. From this assumption one can conclude that the entropy of mixing has the value for an ideal system (e.g., random distribution of atoms A and B on lattice sites of the crystal). However, one may still have $T_0(A) \neq T_0(B)$ and $\Delta H_m(A) \neq \Delta H_m(B)$. To simplify further, still distinguishing between A and B, assume that $\Delta H_m(A)$ and $\Delta H_m(B)$ are constants which are independent of T. Now let

$$R\lambda_A = \Delta H_m(A)[T^{-1} - T_0^{-1}(A)], \tag{8.20}$$

$$R\lambda_B = -\Delta H_m(B)[T^{-1} - T_0^{-1}(B)]. \tag{8.21}$$

With these abbreviations the equations for the liquidus and solidus are:

$$x_B{}^l = [\exp(\lambda_A) - 1]/[\exp(\lambda_A) - \exp(-\lambda_B)], \tag{8.22}$$

$$x_B{}^s = [\exp(\lambda_A) - 1]/[\exp(\lambda_A + \lambda_B) - 1]. \tag{8.23}$$

Of course, the derivation leading to Eqs. (8.22) and (8.23) is oversimplified, but these equations do give a good description [6] of the crystallization curves of Cu–Ni and Si–Ge alloys, as shown in Figs. 8.7 and 8.8.

The calculated solidus curves in Fig. 8.7 and Fig. 8.8 are based on Eqs. (8.22), (8.23), and the experimental values of T_0 and ΔH_m in Cu, Ni and Si, Ge. In both cases the measured solidus curve lies below the calculated one, and the calculated composition gap is smaller than the measured one. A possible explanation for this behavior is the failure to reach thermal equilibrium in the cooling or crystallization process.

REGULAR SOLUTIONS

The simplest way to extend the theory of ideal binary solutions is to add a term to (8.19) corresponding to a heat of mixing of the form

$$\Delta H_{\mathrm{mix}} = \alpha x_{\mathrm{A}} x_{\mathrm{B}} = \alpha x_{\mathrm{A}} (1 - x_{\mathrm{A}}). \tag{8.24}$$

Alloys that are ideal apart from a term of the form (8.24) are said to be strictly regular. More general regular alloys also contain, in addition to the ideal entropy of mixing,

$$S_{\mathrm{ideal}} = -R(x_{\mathrm{A}} \ln x_{\mathrm{A}} + x_{\mathrm{B}} \ln x_{\mathrm{B}}), \tag{8.25}$$

an excess entropy of mixing $T \Delta S_{\mathrm{mix}}$ of the same form as (8.24). In semiconductor solutions $\Delta H_{\mathrm{mix}} - T \Delta S_{\mathrm{mix}}$ is always > 0; i.e., at $T = 0$ the free energy of mixing is always positive. For this reason qualitatively correct results can be obtained for semiconductor alloys by neglecting ΔS_{mix} and retaining ΔH_{mix}, i.e., by treating the semiconductor alloys as strictly regular solutions.

The presence of the term ΔH_{mix} in the free energy means that it costs enthalpy to dissolve one semiconductor in another to form an alloy. However, mixing of the two semiconductors is favored by the ideal entropy term (8.25). At low temperatures the entropy term becomes negligible and phase separation occurs. The critical temperature T_{c} above which the component semiconductors are miscible in all proportions is given by

$$T_{\mathrm{c}} = \alpha/2R, \tag{8.26}$$

where R is the universal gas constant, $R = 2$ cal/mole–deg, or at room temperature (300°K), $RT = 0.6$ kcal/mole $= 25 \times 10^{-3}$ eV per atom pair in an AB crystal. The entire phase-separation curve for a strictly regular solution [7] is sketched in Fig. 8.10.

The examples of Cu–Ni and Ge–Si alloys are almost unique among elemental alloys. The nearly ideal behavior shown in Fig. 8.7 and Fig. 8.8 is found because in each case both elements have the same crystal structure and nearly the same lattice constant (3% difference in Cu–Ni, 4% difference in Ge–Si).

PSEUDOBINARY ALLOYS

When we turn to consider alloys between compound semiconductors, say mixing of $A^N C^{8-N}$ with $B^M D^{8-M}$, we see that the number of possible quaternary alloys of this kind is very large. To simplify the situation, at first let us choose $M = N$ and $D = C$, so that the mixed crystals can be described

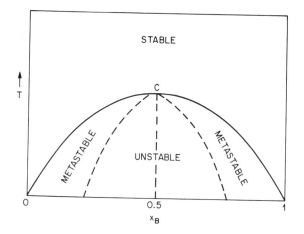

Fig. 8.10. The phase separation curves in the T–x plane for a strictly regular solution [7].

by the simple formula $A_x^N B_{1-x}^N C^{8-N}$ (cation alloy) or $A^N C_x^{8-N} D_{1-x}^{8-N}$ (anion alloy). Mixed compounds of this kind are called pseudobinary alloys. Such mixed compounds have been prepared for a number of systems, especially for the case of cation alloys.

One of the great advantages of pseudobinary alloys is that in thermal equilibrium almost all the cations are located on the cation sublattice, and similarly for the anions, just as would be the case for the parent compounds $A^N C^{8-N}$ and $B^N C^{8-N}$. In the case of a cation pseudobinary alloy $A_x^N B_{1-x}^N C^{8-N}$ the heat of mixing ΔH_{mix} arises from the difference between A–A and B–B interactions in the parent compounds, compared to the A–B interactions that occur in the alloy. But the A and B cations interact only as second-nearest neighbors, separated by intervening nearest-neighbor anions. For this reason ΔH_{mix} is much reduced from the value that it would have in an elemental $A_x B_{1-x}$ alloy. As a result, much wider ranges of miscibility are attainable in $A_x^N B_{1-x}^N C^{8-N}$ alloys than obtain in $A_x B_{1-x}$ alloys.

The next question to consider is whether we can predict the value of T_c for a given pseudobinary alloy $A_x^N B_{1-x}^N C^{8-N}$. By strictly deductive methods this is a hopeless problem, because as we have seen, ΔH_{mix} is of order RT_c, which with $T_c = 1200°\text{K}$ and $x = \frac{1}{2}$ is only 0.025 eV per atom pair. While heats of formation $\Delta H_f°$ are at least of order 1 eV per atom pair, the heats of mixing ΔH_{mix} are 40 times smaller, and by deductive methods no one has yet been able to calculate even $\Delta H_f°$!

The dilemma, thus stated, is insoluble. However, a different statement of the problem leads to a useful result. We have already presented an algebraic formula, Eq. (8.17), which accurately describes $\Delta H_f°(A^N B^{8-N})$.

The formula contains the bond length $d = r_A + r_B$, the covalent and ionic energies E_h and C which depend on d, and on the covalent radii r_A, r_B, and r_C. The formula also contains a metallization factor which in turn depends on the spectroscopic energies E_0, E_1 and E_2. X-ray measurements of a in pseudobinary alloys show that it is quite accurately a linear function of composition x. This suggests that $f_i = C^2/(E_h^2 + C^2)$ also be interpolated linearly in $A_x^N B_{1-x}^N C^{8-N}$ pseudobinary alloys between its values $f_i(AC)$ and $f_i(BC)$ in the parent compounds.

BOWING PARAMETERS

The energies of the gaps labeled E_0, E_1, and E_2 have been measured as functions of composition x, as discussed in later sections of this chapter. The energy E_2, which is similar to the bonding gap E_g, varies slowly and linearly with x. The small gaps E_0 and E_1, which are identified with metallization effects, vary in general more rapidly and may exhibit appreciable non-linearities. One can describe $E_0(x)$ and $E_1(x)$ by the expressions

$$E_{0,1}(A_x^N B_{1-x}^N C^{8-N}) = xE_{0,1}(AC) + (1 - x)E_{0,1}(BC) - \gamma_{0,1}x(1 - x),$$

$$(8.27)$$

where $\gamma_{0,1}$ are the so-called bowing parameters for the $E_{0,1}$ transitions. The bowing parameters γ_0 and γ_1 can also be predicted theoretically in many cases. By putting $x = \frac{1}{2}$ in (8.27) and interpolating on d and on f_i, we can use Eq. (8.17) to calculate $2\Delta H_f^0(A_{1/2}^N B_{1/2}^N C^{8-N})$ and compare this with $\Delta H_f^0(AC) + \Delta H_f^0(BC)$. The difference is equal to $\alpha/2$, and this determines T_c by Eq. (8.26), or more generally the miscibility limits of any $A_x^N B_{1-x}^N C^{8-N}$ alloy system at any temperature. The limits obtained in this way are in generally good agreement with experiment [8]. For example, theory predicts and experiment confirms that at room temperature the following alloys are miscible in all proportions: GaAs–P, In–GaAs, In–GaSb, InAs–P, In–GaP, ZnSe–S, Ag–CuI, CuI–Br, Si–Ge, and GaP–ZnSe. Values of T_c for a number of pseudobinary alloys are given in Table 8.5.

CRYSTALLIZATION OF PSEUDOBINARY ALLOYS

The liquidus and solidus curves of pseudobinary alloys might be expected to show deviations from ideal behavior because of the absence of well-defined cation and anion sublattices in the liquid. However, in practice

TABLE 8.5

Critical Miscibility Temperatures for
Pseudobinary Alloys Which Have Been
Approximated as Strictly Regular
Solutions[a]

Alloy	ΔE [b] (kcal/mole)	T_c [b] (°K)
GaAs–ZnSe	0.65	653
GaP–ZnS	0.41	414
AlSb–ZnTe	0.096	97
InP–ZnSe	1.1	1101
GaAs–ZnS	1.7	1704
InP–ZnS	1.9	1948
GaAs–ZnO	13	12,900
InP–ZnO	14	14,400
GaAs–CdS	0.60	600
InP–CdS	0.31	314
ZnS–Se	0.20	205
ZnS–Te	1.5	1489
ZnSe–Te	0.56	567
Ag–CuI	0.17	166
CuBr–I	0.33	331
InAs–Sb	1.2	1190
GaAs–Sb	0.90	905
GaP–As	0.03	26
Ga–InAs	0.23	226
Ga–InSb	0.27	272
GaP–Sb	1.8	1811

[a] The coefficient α of the heat of mixing
(Eq. 8.24) determines the critical temperature through Eq. (8.26). The mixing heat
parameter α is determined from Eq. (8.17)
using the spectroscopic bowing parameters
Eq. (8.27) as input. Values of α from J. A.
Van Vechten (unpublished).
[b] Here $x = 0.5$.

many pseudobinary alloys still exhibit [6] ideal behavior to within a few
percent in composition x, which represents the limit of experimental errors
even in well equilibrated systems. Crystallization curves for In–GaAs and
InAs–P are shown in Fig. 8.11 and Fig. 8.12 which exhibit nearly ideal
behavior.

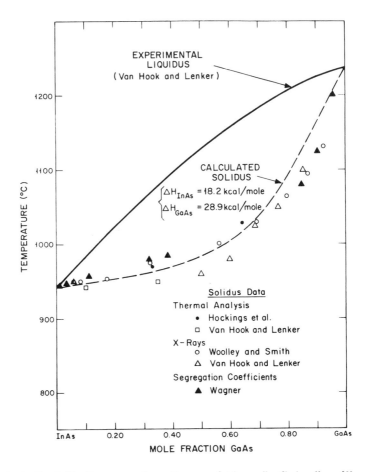

Fig. 8.11. Crystallization curves for cation pseudobinary In–GaAs alloys [6].

VIRTUAL CRYSTAL MODEL

The nearly ideal thermodynamic behavior of pseudobinary $A_x^N B_{1-x}^N C^{8-N}$ alloys is not only surprising from a classical viewpoint: it is even more surprising quantum-mechanically. The true potentials of atoms A and B differ considerably; e.g., the atomic numbers of Ga ($Z_A = 31$) and In ($Z_A = 49$) differ by 50% (see Table 2.1). Moreover, these atoms, with their greatly different potentials, are almost randomly distributed on the cation sublattice. Thus, strictly speaking, we are not dealing in these alloys with periodically repeated potentials, and there is no reason to expect well-defined energy bands with allowed and forbidden energy intervals, or a well-defined energy gap between occupied valence states and empty conduction states.

The very fact, however, that these alloys are thermodynamically almost ideal warns us that once again we have let our physical intuition be led astray by irrelevant mathematics, this time by taking too literally certain quantum-mechanical prescriptions (such as the simplified proofs that derive Bloch functions from translational invariance only). Actually nature is much more flexible and adaptive than schoolbook mathematics, and the properties of pseudobinary alloys are a good illustration of this general point. The alloys in fact have energy bands and energy gaps very

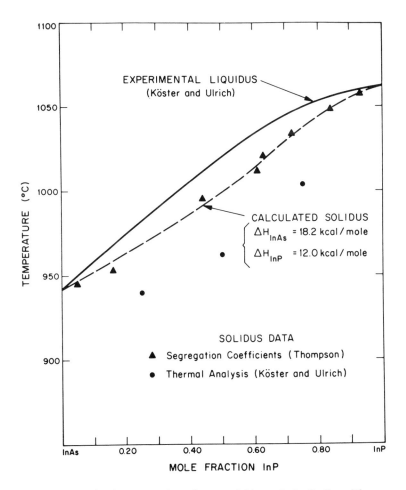

Fig. 8.12. Crystallization curves for anion pseudobinary InAs–P alloys [6].

similar to perfect crystals in many respects. This has led to the so-called *virtual crystal model*, in which the randomly fluctuating cation potential in an $A_x{}^N B_{1-x}^N C^{8-N}$ alloy is replaced by a suitably weighted average cation potential $\bar{V}$ defined by

$$\bar{V}_x = x V_A + (1 - x) V_B. \tag{8.28}$$

Before we apply Eq. (8.28) to study the energy bands of alloys, some caution is needed. For example, in a $Ga_x In_{1-x}P$ alloy, is the appropriate average cation potential at $x = \frac{1}{2}$ that of the element with $Z_A = (31 + 49)/2 = 40$; i.e., are the energy bands of the equimolar alloy $GaInP_2$ those of ZrP? Plainly not, as Ga and In are trivalent nontransition metals, whereas Zr is a transition metal, and ZrP, if it exists, probably does not have the sphalerite crystal structure. What has happened is that we have misinterpreted Eq. (8.28) by averaging the true atomic potential. What we should average is the residual potential seen by the valence electrons outside the atomic cores, which themselves remain intact and, as they are strongly bound to their respective nuclei, are not to be included in the average (8.28).

The proper interpretation of (8.28), then, is that we should average pseudopotentials $V_A{}^p$ and $V_B{}^p$ to obtain an average cation pseudopotential $\bar{V}$. Here some caution is still required. In the alloy the volume Ω_x of the unit cell is different from that of the pure compounds Ω_{AC} and Ω_{BC}. The pseudoatom form factors in Eq. (6.10), however, involve atomic volumes (or unit cell volumes, in the case of compounds). Therefore, the correct form of (8.28) involves the pseudoatom form factors as follows:

$$\bar{V}_x{}^p = (\Omega_{AC}/\Omega_x) x V_A{}^p + (\Omega_{BC}/\Omega_x)(1-x) V_B{}^p. \tag{8.29}$$

With this prescription, one may calculate the energy bands of $A_x{}^N B_{1-x}^N C^{8-N}$ as if it were a pseudocompound with cation pseudopotential $\bar{V}_x{}^p$ and anion pseudopotential V_C. A similar prescription can also be applied to elemental alloys.

OPTICAL TRANSITIONS IN ELEMENTAL ALLOYS

The prototype of semiconductor alloys is the system $Si_x Ge_{1-x}$. Each of the elements has a lowest energy gap which is indirect: although the highest energy in the valence band comes at $\mathbf{k} = \Gamma = 0$, the lowest energy in the conduction band falls elsewhere. In Si it comes near the level X_1 (see Fig. 5.11) along the (100) symmetry axis, whereas in Ge it comes at L_1 (see Fig. 5.12). In $Si_x Ge_{1-x}$ alloys the lowest level in the conduction band remains at L_1 for $0 \lesssim x \lesssim 0.15$, and then switches to the (100) symmetry axis for

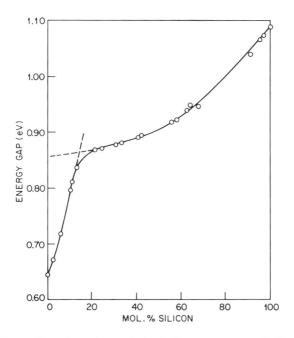

Fig. 8.13. Composition dependence of the indirect energy gap in Ge–Si alloys at room temperature [from R. Braunstein, A. Moore, and F. Herman, *Phys. Rev.* **109**, 695 (1958)].

$x \gtrsim 0.15$. This situation is illustrated by Fig. 8.13, which shows the variation of the optical gap for indirect absorption in the alloy system. The break near $x = 0.15$ is quite clear. The indirect gap in Ge to the (100) minimum can be estimated from these data as 0.86 eV.

At higher energies spectral structure is dominated by the interband edges associated with direct transitions of the kind discussed in Chapter 7. In this case the derivative technique called electroreflectance (see Fig. 7.7) is best used to determine the variation of the direct interband absorption edges with composition x. This variation is shown in Fig. 8.14. To within the limits of experimental accuracy *all* the interband edges appear to vary linearly with x, which does not surprise us greatly, considering the nearly ideal behavior of Ge–Si alloys. In this case the bowing parameters of Eq. (8.27) are zero for all direct transitions to within the limits of experimental uncertainties, even the parameter γ_0 associated with the E_0 transition, which shifts by more than 3 eV on going from Ge to Si. By contrast, the indirect gap $\Gamma \to [(100)\ \text{minimum}]$, which shifts by only 0.25 eV on going from Ge to Si, exhibits a bowing parameter $\gamma_i = 0.4$ eV, according to Fig. 8.13. Thus the indirect gap is much more sensitive to the fluctuating potential in the alloy than are any of the direct gaps. The reason for this is not known, but the result is in accordance with our previous observations [following Eqs. (7.26) and (7.27)] of the difficulties one encounters in trying to predict chemical trends of indirect gaps with the same kind of success achieved for direct gaps.

ENERGY GAPS IN PSEUDOBINARY ALLOYS

The values of the bowing parameters γ_0 and γ_1 for the E_0 and E_1 interband edges have been measured for a number of pseudobinary alloys by electroreflectance. In general γ_0 is larger than γ_1 and can be measured more accurately, as E_0 is the first direct absorption edge in most alloys and

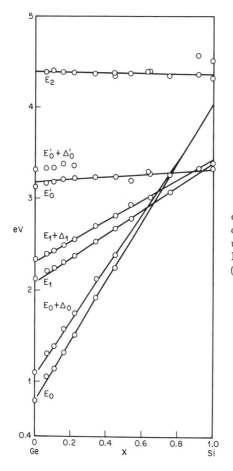

Fig 8 14. Composition dependence of direct energy gap in Ge–Si alloys as determined from electroreflectance measurements [from J. Klein, F. Pollak and M. Cardona, *Helv. Phys. Acta* **41**, 968 (1968)].

TABLE 8.6

Experimental and Theoretical Values of the Bowing
Parameter γ_0 for a Number of Pseudobinary Alloys

	Bowing Parameter γ_0 [a]		
Alloy	DM	EPM	Expt
Gaas–P	0.011	0.014	0.008
InAs–P	0.009	0.008	0.007–0.009
Ga–InSb	0.013	0.009	0.016
Ga–InAs	0.021	0.031	0.012–0.021
InAs–Sb	0.026	0.019	0.021
Ga–InP	0.026	0.050	0.033
ZnS–Se	0.010	0.022	0
ZnSe–Te	0.040	0.033	0.047
ZnS–Te	0.088	0.111	0.089

[a] DM refers to calculations based on the dielectric method, Eq. (7.26), while EPM refers to the empirical pseudopotential form factor method (Chapter 6). The units are 27.2 eV/atom-pair = 626 kcal/mole [from D. Richardson and R. Hill, *J. Phys. C* **6** (1973)].

therefore less subject to corrections due to background absorption. The bowing parameter γ_0 was calculated from Eq. (8.29) with results in generally good agreement with experiment [9]. Experimentally γ_0 is always positive, which means from Eq. (8.27) that the energy bands always bow downward. However, even small errors in defining $\bar{V}$ in the virtual crystal model can reverse the sign of γ_0, so that the calculation of the correct sign and magnitude of γ_0 can be regarded as a severe test of any energy-band model. Theory and experiment are compared in Table 8.6 for the γ_0 bowing parameters.

SUMMARY

Trends in thermochemical properties (heats of formation, melting temperatures, alloy miscibilities) can be explained quantitatively in terms of bond and band parameters. The well-studied $A^N B^{8-N}$ family and its pseudobinary alloys are described quite accurately, and trends in the $A^N B^{10-N}$ family can also be explained. Two factors (ionicity and metallization) are responsible for most of the observed trends.

REFERENCES

1. L. Pauling, "Nature of the Chemical Bonds." Cornell Univ. Press, Ithaca, New York, 1960.
2. R. T. Sanderson, "Chemical Bonds and Bond Energy," Chapter 3. Academic Press, New York, 1971.
3. J. C. Phillips, *J. Phys. Chem. Solids* **34**. (1973).
4. L. C. Snyder, *J. Chem. Phys.* **55**, 95 (1971).
5. J. C. Phillips and J. A. Van Vechten, *Phys. Rev.* **2B**, 2147 (1970).
6. J. Steninger, *J. Appl. Phys.* **41**, 2713 (1970).
7. I. Prigogine and R. Defay, "Chemical Thermodynamics (translated by D. H. Everett), pp. 247–249, 368–370. Wiley, New York, 1954.
8. J. A. Van Vechten, *Proc. Semi. Conf. 10th* p. 602. Nat. Tech. Inform. Service CONF-700801, Nat. Bur. of Std. Springfield, Virginia (1970).
9. D. Richardson and R. Hill, *J. Phys. C* **4**, 339 (1971).

9

Impurities

In earlier chapters we have seen that semiconductors exhibit a number of striking properties. Among the most interesting are: the family of tetrahedrally coordinated $A^N B^{8-N}$ compounds is the largest known family of strictly homologous compounds; many of these crystals are miscible in all proportions at room temperature; and the densities of the crystals are 20% lower than would be found in metallic or ionic structures of the same materials. All of these factors contribute to making it relatively easy to grow single crystals or epitaxial films for technological applications, and they also make it relatively easy to add specific impurities in controlled amounts at predetermined sites in specific geometries. Of course, it is this latter feature which is of the greatest technological value, because it means that we can prepare synthetically crystals with desired electrical properties in specific geometrical configurations. (This is in strong contrast to the situation for most naturally occurring minerals, where precisely controlled doping is still not possible in most cases).

In this chapter I wish to discuss the properties of impurities in semiconductors both in terms of the basic electronic energy levels of the impurity (which are analogous to the energy bands of the host crystal) and also in terms of its interactions with neighboring atoms in the host lattice (which are analogous to the bonds between atoms of the host crystal). The reader will appreciate that the number of possible combinations of impurities and hosts is much larger than the number of hosts alone, and that the electronic properties are more complicated when the impurity atom is not isovalent to the atom it replaces. For this reason my discussion will suffer in accuracy compared to the level maintained in earlier chapters. On the other hand, the reader now has acquired enough background understanding of the properties of the host lattices so that the discussion can be carried out on a realistic basis. Because of the breadth of this subject some of the material given here will be merely factual, whereas in earlier chapters the emphasis has been on matters of principle. This descriptive approach is justified by the general importance of this topic.

CRYSTAL GROWTH AND PERFECTION

Perhaps the most dramatic illustration of the remarkable properties of crystals with the diamond structure is provided by the extremely pure and structurally perfect single crystals grown of Si and Ge. Of course, in thermal equilibrium there will always be a certain concentration of vacancies determined by a Boltzmann factor and by the energy required to form a vacancy. However, the density of macroscopic lattice defects is determined in practice by the history of the sample. When the macroscopic defect results from mismatch of lattice planes (through jogs and slips), one is said to have a dislocation. Dislocation-free crystals of Si and Ge were first grown by Dash. These are probably the most perfect crystals yet prepared [1, 2].†

Once dislocation-free covalent crystals are obtained, they will apparently maintain their perfect structure even when subjected to considerable thermal stress, indicating that in the tetrahedral structure it is not easy to nucleate new dislocations. Dash showed that the thermal stresses generated in a dislocation-free Ge crystal produced by withdrawing it from the melt and then reinserting it did not create any dislocations. On the other hand, when dislocations are already present, they rapidly multiply under stress. For instance, when a crystal containing about 10^3 dislocations/cm² was withdrawn from the melt and then reinserted, the dislocation density increased to more than about 10^6/cm². Today very nearly structurally

---

† See also the introduction by Gatos, the discussion of dislocation-free Si and Ge by W. C. Dash, and the discussion of stoichiometry in compound semiconductors by W. W. Scanlon.

perfect Si crystals several cm in diameter and with very low dislocation densities ($<10^2/$cm^2) are available commercially.

STOICHIOMETRY OF COMPOUND SEMICONDUCTORS

Another illustration of the extraordinary structural stability of tetrahedrally coordinated semiconductors is afforded by the stoichiometric stability of $A^N B^{8-N}$ compounds. In crystals such as GaAs, for example, the presence of a Ga atom on an As site (or vice versa) would create an electrically active impurity which would be readily identified. However, GaAs has been prepared with fewer than 10^{14} active impurities/cm^3, which implies that no more than one atom in 10^9 is situated on the wrong sublattice in very pure GaAs. In other structures (where ionic or metallic binding predominates), as many as half of the cation sites may be vacant, or half again as many cations may occupy interstitial sites. Thus the stoichiometric stability of tetrahedrally coordinated $A^N B^{8-N}$ compounds is a very striking confirmation of the strength of the sp^3 covalent bond. Incidentally, the compound in this family which is known to exhibit greatest deviations from stoichiometry is AgI, which contains many Ag vacancies. Note that the ionicity of AgI is within 2% of the critical value (Table 2.2), and that it transforms under very little pressure to the NaCl structure (Fig. 8.5). Another family of materials which is not stoichiometrically stable is the PbS family, which may contain either excess cations or excess anions [3]. Presumably this instability is nurtured by the high polarizability of the nonbonding s electrons. We saw in the last chapter that these nonbonding electrons also give rise to the unusual heats of formation in this family (Fig. 8.2).

SHALLOW AND DEEP IMPURITY STATES

The impurity levels which have been studied most extensively are those with small binding energies of order $k_B T$ at $T = 300°$K, i.e., those with binding energies of order 0.025 eV or less. These are referred to as shallow impurity states. Deep impurity states have binding energies of 0.1 eV or more, and the energy levels associated with them lie towards the middle of the minimum semiconductor energy gap ΔE_{cv}. The shallow states are the ones for which electrical activity can be controlled by temperature and voltage, and generally speaking their concentration can also be controlled over a wider range. The impurities associated with deep states, on the other hand, disturb the bonds of the host crystal so greatly that their solubility is much reduced. Such deep states are sometimes called traps.

DIFFUSION OF INTERSTITIAL AND SUBSTITUTIONAL IMPURITIES

Most of the shallow impurity states which we shall encounter are based on substitution by the impurity atom for one of the host atoms. In spite of the apparently open nature of the tetrahedrally coordinated structures, an atom or ion must be very small to occupy an interstitial site. In fact, appreciable numbers of interstitials have been demonstrated in Si and Ge only for atomic H and He, and for Li^+ ions. Only the Li is electrically active. Diffusion studies have shown that Li diffuses approximately twice as rapidly in Ge and Si as does He, but in Si hydrogen appears to diffuse more rapidly than Li. The thermal activation energies for diffusion of Li in Ge and Si are 0.5 and 0.7 eV, respectively.

The remaining impurities which enter $A^N B^{8-N}$ semiconductors almost certainly take up substitutional sites; i.e., they replace atoms of the host lattice. Such atoms diffuse into the lattice by exchanging lattice sites with vacancies. The activation energies [3] for diffusion of columns III or V impurities in Si and Ge are typically: (III, Si) : 3.5 eV; (V, Si) : 3 eV; (III, Ge) : 3 eV; and (V, Ge) : 2.5 eV. The similarities of the activation energies of elements from the same column suggests that atomic radii do not play an important role in the activation process, and that the important energy is an electrostatic one involving both the interaction between the vacancy and the impurity, as well as the polarization (or deformation) of the valence bonds of the host lattice in the diffusion process. This picture explains the generally higher activation energies in Si (where the cohesive energy is about 20% greater than in Ge; see Table 8.1). Moreover, vacancies probably carry a net negative charge, because of spillover of electrons from adjacent atoms. This causes the column V impurities to associate with the vacancies, because column V impurities act as donors (see below) and hence are positively charged. The long-range Coulomb interaction brings the donors closer to the vacancies, and leads to a lower apparent activation energy, in agreement with the values quoted above. Similarly, column III impurities act as acceptors and carry a net negative charge. For this reason at intermediate distances they are weakly repelled by negatively charged vacancies, and hence appear to have a higher activation energy for diffusion.

DISTRIBUTION COEFFICIENTS

There are several ways of growing crystals. Most of them involve equilibrium (or near equilibrium) between the solid and the liquid (or melt).

Crystals can also be grown by vapor deposition techniques. For purification of crystals one takes advantage of the fact that almost all impurities prefer to be in the melt rather than in the crystal. The liquid, of course, is flexible so far as density and coordination number of each atom are concerned, and so liquid Si or Ge readily accommodates impurities of various kinds. But the tetrahedrally coordinated semiconductor crystal has a very special density and a very special coordination configuration which does not accommodate impurities at all readily. This difference in distribution of a component between liquid and solid phases in equilibrium is described in terms of the distribution coefficient

$$k = x_S/x_L, \tag{9.1}$$

where x_S and x_L represent the concentrations of impurities in the solid and liquid, respectively. In thermodynamic equilibrium k is indeed a constant for x_S and $x_L \ll 1$; it coincides with the ratio of what are called the activity coefficients of the impurity in the liquid and in the solid.

As one would expect from our qualitative discussion, values of k are very small for tetrahedrally coordinated semiconductors. Even near the melting point (1400°C), the value of k for Zn impurities in Si is only 10^{-5}. The only impurities which dissolve appreciably in Ge and Si are ones from columns III, IV, or V of the periodic table. Curves of $k(T_m)$, where T_m is the melting point, plotted against the covalent radius of the impurity, are shown in Fig. 9.1 for impurities in Ge, and in Fig. 9.2 for impurities in Si. According to these figures, the distribution curves fall into groups indexed by the columns of the periodic table. As one would expect, the isovalent column IV elements dissolve most readily in the column IV host crystals. The difference between the column IV curve and the average of the columns III and V curves is about a factor of e^4 in Ge and a factor of e^3 in Si. With $T_m = 1200°$K in Ge and 1700°K in Si, the corresponding difference in activation energies [derived from assuming $k(T_m) \propto \exp(-\Delta H/k_B T_m)$] are about 0.5 eV, or about $\Delta E_{cv}/2$. This is roughly what one would expect, on the grounds that adding a column III or column V impurity to Si or Ge requires adding an electron not at the Fermi energy but at an energy $\Delta E_{cv}/2$ away from the Fermi energy.

There are many aspects of distribution coefficients which are not understood. For example, in Ge $k(\text{III}) > k(\text{V})$, but in Si, $k(\text{V}) > k(\text{III})$, as one can see from comparing Fig. 9.1 with Fig. 9.2. The lattice constants of Ge and Si differ by only 4%, and it seems that this reversal cannot be explained by the difference in lattice constants. The largest difference between Si and Ge is that Si has no d core electrons, while Ge does, but it is not clear how this could account for the reversal of the distribution coefficients.

DONORS AND ACCEPTORS

When an impurity from column V of the periodic table (N, P, As, Sb, or Bi) substitutes for a Si atom in a silicon crystal, one may picture the distribution of its five s or p valence electrons as follows. Four of them enter the valence band, replacing the four valence electrons of the missing Si atom. According to the exclusion principle, the fifth electron must go into conduction band states, because all the valence band states are occupied. The fifth electron is still attracted by the residual charge of $+e$ on the column V impurity, and this leads to a small binding energy which will be discussed below. The energy is sufficiently small that at room temperature there is a good probability that the electron will be ionized from the impurity, and so be free to carry current. The column V impurity has donated one electron to the conduction band of the crystal, and it is called a monovalent donor.

The situation with column III impurities is similar in many respects. Here the impurity lacks one electron to complete its valence bonds. One may imagine that an electron is removed from valence bonds (or the valence band) elsewhere in the crystal to complete the bonds around the impurity.

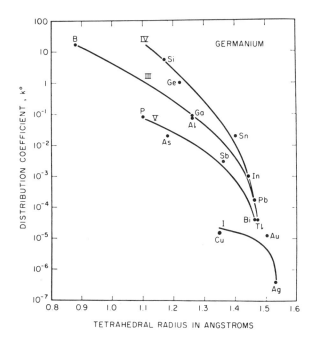

Fig. 9.1. Distribution coefficients $k(T_m)$ at the melting point T_m of germanium as a function of tetrahedral radii [from F. A. Trumbore, *Bell Syst. Tech. J.* **39**, 220 (1960)].

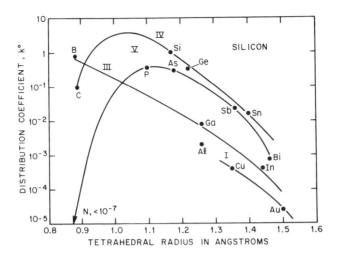

Fig. 9.2. Distribution coefficients $k(T_m)$ at the melting point T_m of silicon as a function of tetrahedral radii [from F. A. Trumbore, *Bell Syst. Tech. J.* **39**, 221 (1960)]. The data point for C is more recent [from A. R. Bean and R. C. Newman, *J. Phys. Chem. Solids* **32**, 1211 (1971)]. Note that although B lies on the extrapolated curve for column III elements, C lies well below the extrapolated column IV curve, as does N for the column V curve (est. value of $k < 10^{-7}$). This may be explained through formation of multiple (short) bonds in the melt by C and by N. See also the discussion.

The missing electron is effectively positively charged, and is called a hole. The impurity is negatively charged, because it has accepted a valence electron from the valence band of the crystal. Thus the group III impurity is called an acceptor. The negative charge of the acceptor attracts the positively charged hole, which leads to weak binding. The positively charged hole may be ionized, and it is then available to carry an electric current. Electrons and holes are distinguishable, of course, in a magnetic field through the transverse Hall effect.

A schematic survey of the energy levels of donor and acceptors is shown in Fig. 9.3 for Ge. The shallow donors are shown on the left near the bottom of the conduction band, while the shallow acceptors are just above the top of the valence band, also on the left. In a crystal with equal numbers of shallow donors and acceptors the Fermi energy will fall almost at the middle of the gap, as indicated by the dashed line. The various states with different possible charges shown towards the middle of the gap are the trap states, which are ordinarily negligible in number for well-purified crystals.

An approximate energy-level scheme for donors and acceptors in Si is shown in Fig. 9.4. The primary difference from Ge is that all the binding energies are much larger, for reasons discussed below.

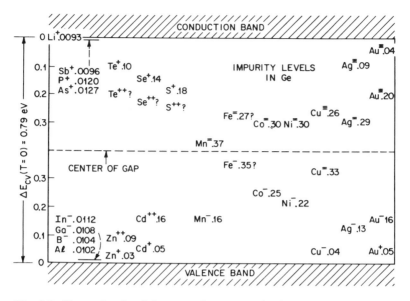

Fig. 9.3. Energy levels of donors and acceptors in Ge. The shallow monovalent donors and acceptors are on the left of the diagram, the polyvalent and transition metal levels are deeper (trap states).

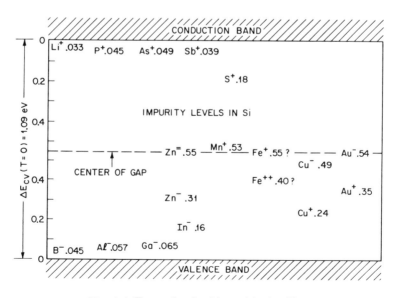

Fig. 9.4. Energy levels of impurities in silicon.

ISOVALENT IMPURITIES

In compound semiconductors (such as GaP) it has recently been found that even isovalent atoms may act as impurities [4]. For example, when N replaces P in GaP, the electronegativity of the N atom is so great that it attracts an extra valence electron to itself, and becomes negatively charged. This makes N an isovalent acceptor in GaP. Of course, the Coulomb field of the negatively charged N atom can weakly bind a hole, as discussed below, and this hole can be ionized to make it available to carry current. One should imagine the additional electron as being very tightly bound to the N atom, while the additional hole circles around the impurity at distances of several lattice constants. The opposite situation, in which a P atom is replaced by a much less electronegative Bi atom, gives rise to a tightly bound hole and a weakly bound electron. Thus Bi is an isovalent donor in GaP.

In the literature isovalent impurities are often called isoelectronic impurities. This nomenclature, although by now well accepted for this situation, is a misnomer from the point of view of general chemical usage. The compounds BN and BeO are isoelectronic in the sense that core and valence electron configurations in both cases are $(1s^2)_2 2s^2 2p^6$. On the other hand, N and Bi are isovalent to P, because all three atoms belong to the same column of the periodic table, although the core electron configurations are all different.

SPHERICAL (HYDROGENIC) MODELS

The shallow impurity states associated with monovalent donors and acceptors give rise to weakly bound states of electrons and holes. The energy levels of these impurities can be measured with great precision in many cases by preparing very pure host crystals and then doping the crystals with a very low concentration of the desired impurity (at levels of order 10^{-6} impurities per host atom). The energy levels are then studied in the infrared by the same methods used to analyze the visible and ultra-violet spectra of one-electron atoms or ions in gaseous states. The role played by the vacuum for the atom in the gaseous state is played by the host crystal itself in the case of the impurity.

What are the differences between atoms in vacuum and impurity atoms in the crystal? The most important difference is that the crystal is polarizable, so that at long distances the Coulomb interaction in vacuum

$$V_{\text{Coulomb}} = -e^2/r \qquad (9.2)$$

is replaced by the modified interaction

$$V_{\text{Coulomb}} = -e^2/\bar{\epsilon}r, \tag{9.3}$$

where $\bar{\epsilon}$ is the dielectric constant which should describe the reduction in field between the extra particle and the impurity in the crystal. We shall return to the microscopic significance of $\bar{\epsilon}$ below. For the moment it is enough to notice that (9.3) is equivalent to (9.2) if we replace e^2 in (2) by $e^2/\bar{\epsilon}$.

The second difference concerns the kinetic energy of the extra particle (electron or hole) in the crystal. In general, kinetic energy is described in terms of the relationship between the energy of the particle in the absence of the Coulomb field of the ion or impurity, and the momentum of the particle. For an electron in vacuum the kinetic energy is simply

$$E_{\text{kinetic}} = E_0 + p^2/2m, \tag{9.4}$$

where E_0 is the zero of energy and m is the electron mass. In the crystal the extra particle should be thought of as a wave packet which travels with the usual group velocity

$$\mathbf{v} = (1/\hbar)(\partial E/\partial \mathbf{k}). \tag{9.5}$$

The relationship (9.5) reduces to $\mathbf{p} = m\mathbf{v}$ if E is replaced by E_{kinetic} from (9.4) and $\mathbf{p}$ is identified with $\hbar\mathbf{k}$.

Near the bottom of the conduction band or the top of the valence band, the relationship between energy and momentum is still quadratic, as in (9.4), providing we are concerned with energies small compared to the energy gaps in the crystal. This condition is satisfied in practice for shallow donors and acceptors (binding energies of order 0.01 eV compared to energy gaps of order 1 eV in Si and Ge). However, the relationship between E and $\mathbf{p} = \hbar\mathbf{k}$ is more complicated. First of all, if the band edge is not located at $\mathbf{k} = 0$, but instead is located at $\mathbf{k} = \mathbf{k}_0$, then E will be quadratic in $(\mathbf{k} - \mathbf{k}_0)^2$, or in $(\mathbf{p} - \mathbf{p}_0)^2$, where $\mathbf{p}_0 = \hbar\mathbf{k}_0$. Because of crystal symmetry, however, the quadratic terms need not be isotropic; i.e., the coefficients of the x, y, and z components need not be equal. Finally, allowances must be made for the effects of degeneracies, as discussed below. The values of the coefficients of $(\mathbf{k} - \mathbf{k}_0)^2$ can be expressed in units of the electron mass and are described as effective masses. We have already discussed these quantities in Chapter 5; see especially Tables 5.1 and 5.2.

One approach to describing the kinetic energy of the wavepacket that is useful for qualitative purposes is to continue to use (9.4), but to replace m by m^*, where $(m^*)^{-1}$ is a suitably weighted average of the coefficients of $(\mathbf{p} - \mathbf{p}_0)^2/2$. One weighting which suggests itself, for example, is the one which gives rise to the same density of states $N(E)$ per unit energy E.

This weighting in the case of anisotropy is equivalent to taking an harmonic average and it is consistent with phase-space considerations. In the following qualitative discussion, this is the procedure that is adopted. A more quantitative approach, based on the exact form of $E_{kinetic}$ required by the known form of the energy bands, can be based on trial wave functions which are anisotropic in the same way that $E_{kinetic}$ is. This variational approach has been discussed by Kohn [5].

The extra particle that interacts with the impurity now has the energy of an electron bound to a proton, but with e^2 replaced by $e^2/\bar{\epsilon}$ and m replaced by m^*. This gives rise to a hydrogenic set of energy levels, with the binding energy of the 1s ground state given by

$$E_{1s} = (m^*e^4/2\hbar^2\bar{\epsilon}^2) = (m^*/m)(1/\bar{\epsilon}^2) \text{ Ry.} \qquad (9.6)$$

Typical values for m^*/m are 0.3 for electrons in Si and 0.1 for electrons in Ge. The appropriate values for $\bar{\epsilon}$ are 11.7 in Si and 16.0 in Ge. Substituting these values in (6) and replacing 1 Ry by 13.6 eV, we obtain $E_{1s}(\text{Si}) = 0.03$ eV and $E_{1s}(\text{Ge}) = 0.01$ eV, in approximate agreement with the binding energies for shallow donors shown in Figs. 9.3 and 9.4.

From equation (9.6) it is clear that the factor $\bar{\epsilon}$ plays an important role in determining the binding energy of the impurity even in the simplified model. In Chapter 4 we discussed the dielectric polarizability at low frequencies and saw that this could be described by two dielectric constants, ϵ_0 and ϵ_s. If we consider the ions which make up the lattice to be fixed at their respective lattice sites, then only the valence electrons are polarizable. Their contribution to the dielectric constant is represented by ϵ_0, which is the low frequency limit of the optical dielectric constant $\epsilon_1(\omega) + i\epsilon_2(\omega)$ described by Eqs. (7.2) and (7.3). [For $\hbar\omega < \Delta E_{cv}$, $\epsilon_2(\omega) = 0$ and as $\omega \to 0$, $\epsilon_1(\omega) \to \epsilon_0$].

The second contribution to the crystal polarizability is that of the ions, and it is given by $\epsilon_s - \epsilon_0$. This contribution becomes effective only at frequencies near or below the frequency ω_{t0} of transverse optic vibrations [see Eq. (4.9)]. In the frequency range $\omega_{t0} < \omega < \Delta E_{cv}/\hbar$, $\bar{\epsilon}$ is approximately given by ϵ_0, while for $\omega < \omega_{t0}$, $\bar{\epsilon}$ is given by ϵ_s. The physical picture of the ion polarizability requires each ion to carry some net charge. For Si and Ge, however, $e^* = 0$, and according to Eqs. (4.11) and (4.12), this means that $\epsilon_s = \epsilon_0 = \bar{\epsilon}$ in all cases.

The ambiguity, then, in $\bar{\epsilon}$ arises for impurities in compound, partially ionic semiconductors such as GaAs or ZnSe. In the crystals the electric fields between extra particles and the impurities to which they are bound are screened by electronic polarization (ϵ_0) and by part of the lattice polarization, i.e., $\epsilon_0 < \bar{\epsilon} < \epsilon_s$. To the extent that the motion of the particle around the impurity is a slow one, the ions will follow the particle and

$\bar{\epsilon} = \epsilon_s$. To the extent that it is fast the ions will not be able to follow the particle and $\bar{\epsilon} = \epsilon_0$. Whether the motion is fast or slow depends on the magnitude of the binding energy E_{1s} compared to the transverse optic mode frequency $\hbar\omega_{t0}$. An explicit formula [6] for $\bar{\epsilon}$ is

$$\bar{\epsilon}^{-1} = \epsilon_s^{-1} + 4(\epsilon_s^{-1} - \epsilon_s^{-1})(E_{1s}/\hbar\omega_{t0})A, \qquad (9.7)$$

$$A = \ln\tfrac{1}{2}[1 + (1 + \hbar\omega_{t0}/E_{1s})^{1/2}].$$

At this point it is helpful to summarize the virtues and deficiencies of the hydrogenic model for impurity binding energies, as represented by Eqs. (9.6) and (9.7). The chief virtue of the potential (9.3) is its spherical or isotropic character, and this applies to the kinetic energy (9.4) as well. The deficiencies of the model are:

i. As we have already mentioned, because of the cubic symmetry of the crystal, the kinetic energy operator is usually more complicated than (9.4), and in particular there may be degeneracies in the crystalline energy levels that are not contained in the hydrogenic model. These degeneracies are discussed in the following two sections.

ii. In the unit cell containing the impurity, the potential cannot have the form (9.3) and certainly must vary from impurity to impurity, just as the experimental binding energies vary. These variations are called chemical shifts, and they will also be discussed in a later section.

BAND-EDGE DEGENERACIES

If the maximum energy in the valence band or the minimum energy in the conduction band were to occur in a single band at $\mathbf{k} = 0$, then because of the cubic symmetry of the crystal, the kinetic energy of a particle wave packet made up out of states from the vicinity of this band edge would necessarily be described by Eq. (9.4). This situation actually does hold for the conduction band edges in a number of III–V and II–VI semiconductors such as GaAs, InSb, and CdTe. However, the valence band edge, although it falls at $\mathbf{k} = 0$, is always degenerate and the conduction band edges in Si, Ge, and GaP do not fall at $\mathbf{k} = 0$. This gives rise to degeneracies which modify and to some extent complicate the spectra of monovalent donor and acceptor shallow impurities.

Consider the conduction band edges in Ge first. These are centered on the points $\mathbf{k} = \mathbf{L}$ in Fig. 4.1, which are the centers of the $(1, 1, 1)$ faces of the Brillouin zone. The edges are nondegenerate and with $\mathbf{k}' = \mathbf{k} - \mathbf{L}$ the energy is described by

$$E(\mathbf{k}) = E(\mathbf{L}) + \hbar^2[(k_l')^2/2m_l + (k_t')^2/2m_t]. \qquad (9.8)$$

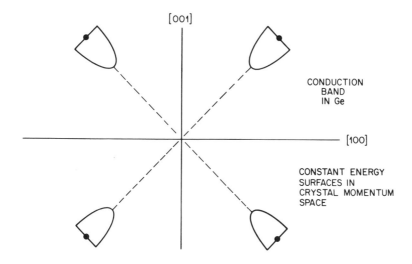

Fig. 9.5. Ellipsoidal surfaces of constant energy, plotted in momentum space for Ge. The energy is slightly greater than that at the bottoms of the conduction band. Ellipsoidal principal axes lie in $(1, 1, 1)$ directions, and the surfaces have been projected onto the xz plane in the figure.

The surfaces of constant energy for E slightly greater than $E(\mathbf{L})$ are ellipsoids of revolution about the axis ΓL, as shown in Fig. 9.5. The ellipsoids are quite anisotropic, for as shown in Table 5.2, m_l/m_t is about 20. There are four inequivalent complete ellipsoids, those at $(1, 1, 1)$ being equivalent to those at $(-1, -1, -1)$.

The situation in Si is similar to that in Ge, except that now the ellipsoids are centered on points near but not at the X or $(2, 0, 0)$ faces of the Brillouin zone. Also the ellipsoids are less anisotropic, the axial ratio m_l/m_t being about five, according to Table 5.2. Surfaces of constant energy are shown in Fig. 9.6.

The effective masses m_t and m_l were first measured in elegant cyclotron resonance experiments [7]. In an external magnetic field the motion of a particle wavepacket is described by the usual Lorentz equation

$$dp/dt = (e/c)\mathbf{v} \times \mathbf{H}, \qquad (9.9)$$

where $\mathbf{v}$ is the group velocity of the wave packet. This is given by

$$\mathbf{v} = (\hbar)^{-1}\nabla_k E(\mathbf{k}). \qquad (9.10)$$

When the effective mass is anisotropic, so that $E(\mathbf{k})$ is given by (9.8), the cyclotron resonance frequency can be calculated easily by choosing coordinate axes with the z axis parallel to the principal axis of (9.8), either

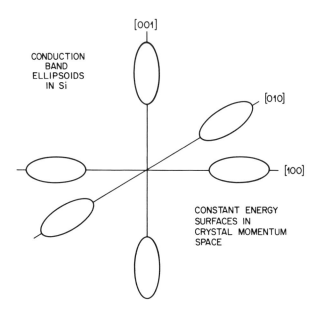

Fig. 9.6. Ellipsoidal surfaces of constant energy, plotted in momentum space for Si. Again the energy is slightly greater than that of the conduction band edge.

ΓL (Ge) or ΓX (Si). If the magnetic field lies in the xz plane at an angle θ with the z axis, then the electrons follow elliptical orbits about the magnetic field at the cyclotron resonance frequency ω_c. This can be calculated from (9.9) by letting $\mathbf{p}$ have the usual time dependence $\exp(i\omega t)$ and separating (9.9) into components. The result is

$$\omega_c = eH/m^*c \tag{9.11}$$

with the effective mass m^* having the ellipsoidal form

$$(1/m^*)^2 = (\cos^2\theta)/m_t^2 + (\sin^2\theta)/m_t m_l. \tag{9.12}$$

Thus a measurement of ω_c as a function of θ determines both m_t and m_l. (It can also be used to orient the crystal without benefit of X-ray measurements). The pattern of resonance frequencies of electrons in Ge is shown in Fig. 9.7, while the resonance pattern for electrons in Si is shown in Fig. 9.8. The values of the effective masses m_l and m_t in Ge and Si quoted in Table 5.2 were obtained from more recent cyclotron resonance experiments.

Cyclotron resonance experiments have also been performed on holes in Si and Ge. For the holes the description and interpretation of the resonance frequencies is complicated by the orbital degeneracies of the top of the valence band. These degeneracies are treated by matrix methods, leading eventually to certain band parameters of the kind described in Chapter 5.

The band parameters deduced in this way are in good agreement with those calculated by the pseudopotential method, using energy bands derived from study of the fundamental optical spectra of the crystals (Chapter 7). In this way the validity of the description of particle kinematics in terms of wave packet motion is confirmed both qualitatively (in terms of crystal symmetry) and quantitatively (in terms of one-electron energy bands). One can therefore proceed with confidence to apply the method of wave packets to the study of impurity levels. We describe the results only for shallow donors, because of the simplicity of equation (9.8), but more elaborate matrix calculations have been successful for shallow acceptors as well.

VALLEY ANISOTROPIES

To describe donor states made up out of wave packets from energy surfaces such as those shown in Figs. 9.5 and 9.6, one uses the principle of

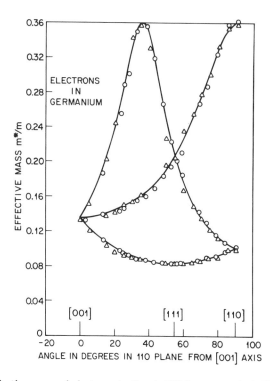

Fig. 9.7. Effective mass of electrons in Ge at $4°K$ for magnetic field directions in a (110) plane. The curves are calculated from Eq. (9.12) by choosing $m_l/m = 1.58$ and $m_t/m = 0.082$.

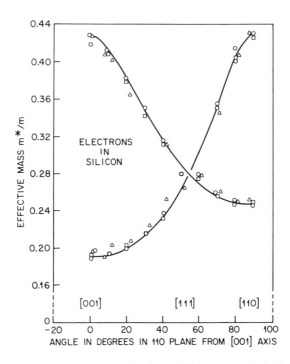

Fig. 9.8. Effective mass of electrons in Si at 4°K for magnetic field directions in a (110) plane. The theoretical curves are calculated from Eq. (9.12) by choosing $m_l/m = 0.98$ and $m_t/m = 0.19$.

superposition. To be specific consider donor states in Ge. The nl hydrogenic state becomes $\psi_{nl\alpha}$, which has the form of a sum over valleys labelled by j,

$$\psi_{nl\alpha}(\mathbf{r}) = \sum_{j=1}^{4} \alpha_j \varphi^j_{n\,l}(\mathbf{r}), \qquad (9.13)$$

where $\varphi^j_{n\,l}(\mathbf{r})$ is a one-valley wave function, derived from a Schrödinger equation based on the kinetic energy given by Eq. (9.8).

The implications of the valley-sum (9.13) for different hydrogenic states can be illustrated by examples. When the kinetic energy is given simply by $p^2/2m$, the 1s ground state wave function ψ_{1s} is proportional to $\exp(-r/a_0)$, where $a_0 = \hbar^2/e^2m$. But when the kinetic energy has the anisotropic form (9.8), then φ^j_{1s} is also anisotropic and is approximately proportional to $\exp\{-[z^2/a_l^2 + (x^2 + y^2)/a_t^2]^{1/2}\}$, where $a_l = \hbar^2/e^2m_l$ and $a_t = \hbar^2/e^2m_t$. (Note that for each valley the z axis is chosen along the principal axis of revolution of that valley, so that the anisotropy of φ^j_{1s} is actually different for each j).

For orbitally degenerate states with $l \geq 1$, the cylindrical anisotropy of the kinetic energy described by equation (9.8) gives rise to axial splittings of states with different values of $|m_l|$, where m_l is the azimuthal quantum number relative to the principal axis of the kinetic energy ellipsoid of revolution. These splittings are illustrated in Fig. 9.9 for the 2p and 3p donor states of Ge and Si. The splittings are relatively greater in the case of Ge because the mass anisotropy is greater. In Ge the axial splitting is so large that the $3p_0$ level lies below the $2p_{\pm 1}$ levels.

Experimentally it is found that the energy levels of different donors (interstitial Li, substitutional P, As, Sb, and Bi) agree very well with theory [8] for $l \geq 1$. But for the s states there are larger discrepancies (especially for the 1s state), and these are found to vary from one donor to another. Such variations are called chemical shifts, because they are specific to the element that acts as donor. Values of binding energies of shallow donors and acceptors in Si and Ge are listed in Table 9.1.

CHEMICAL SHIFTS AND CENTRAL CELL CORRECTIONS

An important consequence of the shallowness of the binding energies of monovalent donors and acceptors in Si and Ge is the corresponding increase in orbital radius of the bound states. In the hydrogenic model the

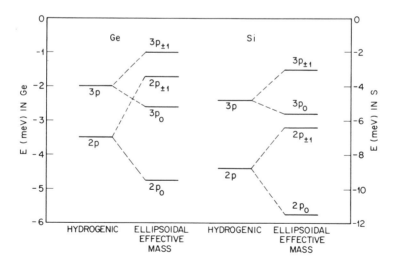

Fig. 9.9. The splitting of hydrogenic p levels caused by ellipsoidal kinetic energy anisotropy. The values shown are realistic ones for donors in Ge and Si. Note the crossing of 2p and 3p levels in Ge, where the mass anisotropy is large.

TABLE 9.1

Binding Energies of Shallow Donors and Acceptors
in Si and Ge[a]

Donor	Si	Ge	Ac-ceptor	Si	Ge
P	45.8	12.9	B	44.4	10.47
As	53.8	14.2	Al	68.9	10.80
Sb	43.1	10.3	Ga	72.7	10.97
Bi	—	12.8	In	156.2	11.61
			Tl	—	13.10

[a] The energies are in units of 10^{-3} eV, and the data
were collected by Baldereschi and Hopfield [4].

Bohr radius of the 1s state is derived by scaling arguments [cf. Eq. (9.6)]:

$$a_{1s} = \hbar^2 \bar{\epsilon}/m^* e^2 = (m/m^*) \bar{\epsilon} a_0, \qquad (9.14)$$

where $a_0 = \hbar^2/me^2$ is the Bohr radius of the hydrogen atom. In Ge the
effective Bohr radius from (9.14) is greater than that of the hydrogen atom
by a factor of 150. This means that most of the donor states spend very
little time in the atomic cell containing the impurity. The states with $l \geq 1$
spend very little time even in cells adjacent to the impurity (which may
have been strained somewhat). This explains why the effective mass theory
and experiment agree so well for these states.

For the 1s state the situation is different. Here we must consider the
effect on the energy $E_{1s\alpha}$ of the state $\Psi_{1s\alpha}$ of choosing different values of α_j
in equation (9.13). Each wavepacket $\Phi_{nl}^j(r)$ from the jth valley is proportional to the hydrogenic envelope function just described, as well as to the
Bloch function at the jth band edge. Thus the general form of $\Phi_{nl}^j(r)$ is that
of a rapidly oscillating Bloch function amplitude-modulated by an anisotropic hydrogenic function. Interference terms between Φ_{nl}^j and $\Phi_{nl}^{j'}$ are
small in most cases, because the j and j' Bloch functions interfere destruc-
tively. However, in the impurity cell itself the potential is very much
stronger than in other cells of the crystal. In this cell the 1s envelope function also has a large amplitude. Therefore, if we choose the coefficients α_j
so that the sum in (9.13) has its maximum value for $\mathbf{r} = 0$, we will get a
lower energy than we would have gotten with any other choice. In fact, if
we choose the phases of $\varphi_{nl}^j(\mathbf{r})$ so that $\Phi_{nl}^j(0)$ is the same for all j, then α_j is
the same, independent of j, in the state of lowest energy. This state is totally
symmetric (s-like symmetry) in the α_j's, as well as having 1s-like envelope
functions in the one-valley functions Φ_{nl}^j. One can show that there are

three other allowed combinations of the α_j's, and that these have p-like symmetry, and so have zero amplitude in the impurity cell. Thus their energy is higher than that of the totally symmetric state discussed above.

The splitting that we have just described arises from the degeneracy of the valleys shown in Figs. 9.5 and 9.6, and for that reason it is called a valley–orbit splitting, because it involves the orbital functions $\Phi_{1s}^j(r)$ from different valleys. The effects of the degeneracy and the valley-orbit splitting on the 1s donor ground states of Ge and Si are shown in Fig. 9.10. For substitutional monovalent donors (P, As, Sb, Bi) the ground state is the nondegenerate, totally symmetric combination of valleys. In Si there are five other combinations of the α's, one with p-like symmetry and threefold degenerate, and one with d-like symmetry and doubly degenerate. In Ge there are only three other combinations of the α's and these form a triply-degenerate, p-like state. Only the totally symmetric combination has an energy which differs greatly from that predicted by effective mass theory.

An interesting aspect of the central cell corrections is that only for Li impurities is the energy of the totally symmetric combination of valleys greater than that of the other combinations. This reflects the weakness of the Li$^+$ potential compared to that of the Si atoms and the fact that the Li$^+$ atom is not located on a lattice site.

The qualitative explanation for the chemical shifts of impurities is that they arise from differences in electronegativity between the substitutional impurity and the host atom which it has replaced. According to this picture, with the electronegativities taken from Table 2.5, the smallest relative

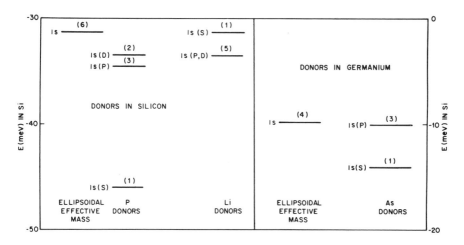

Fig. 9.10. Some valley–orbit splittings of 1s ground states of donor impurities in Si and Ge [8].

correction should be associated with Sb impurities in Ge, followed by Sb impurities in Si. This is found to be the case experimentally.

The isovalent impurities (such as N substituting for P in GaP) represent an extreme case in which the attractive potential that binds the first particle (in this case, an electron to a N atom) is short-range in character, and there is no Coulomb potential at all. Whereas the Coulomb field of monovalent acceptors or donors always binds particles, leading to an imbalance of electronic charge in the impurity cell, a certain minimum difference in electronegativity must be exceeded before the short-range potential of an isovalent impurity can bind particles. This difference must be calculated after making allowance for the effects of lattice relaxation in the neighborhood of the impurity. When this is done, the threshold for binding can be predicted [8] from equations such as (7.24).

IMPURITY STATES IN COMPOUND SEMICONDUCTORS

The number of possible impurity configurations is doubled when one considers a compound semiconductor (such as GaP) instead of an elemental semiconductor (such as Si). In III–V semiconductors, for example, elements from column II may substitute on the cation lattice as acceptors, while elements from column VI may substitute on the anion sublattice as donors. Elements from column IV act as donors on the cation sublattice and as acceptors on the anion sublattice.

Apart from GaP and AlP all the common compound semiconductors have their conduction-band minima at $\mathbf{k} = 0$. This minimum is non-degenerate and is labeled by Γ_{1c} (see Fig. 5.10). The effective mass m_c^*/m at this minimum is quite small in groups IV and III–V semiconductors, and is still small (~ 0.2) even in II–VI semiconductors such as ZnSe (see Table 5.1). According to equation (9.6), this means that the binding energy associated with monovalent donors is small in all compound semiconductors except GaP. For that crystal the effective masses are similar to Si, and have the same (100) ellipsoidal symmetry (Table 5.2). Because the dielective constant is smaller, the effective mass-binding energies are about twice as large for donors in GaP as for donors in Si.

The binding energies of acceptors are in some respects more interesting. The top of the valence band occurs at $\mathbf{k} = 0$ in all cases, and the effective mass parameters lead to an equivalent density-of-states mass m^* which is about 0.3–$0.5m$ in all cases (Lawaetz, cited in Chapter 5). Nevertheless, large variations in acceptor-binding energies are observed in compound semiconductors. The origin of these variations is the same as that responsible for binding of carriers to isovalent impurities, that is, electronegativity differences which shift the average energy in the unit cell containing the impurity [Eq. (7.24)].

The distribution coefficients for various impurities are not yet well understood for most compound semiconductors. Moreover, chemical identification of the impurities in most materials is uncertain. Two compound semiconductors which have been studied in detail are GaP and CdS. Binding energies of donors and acceptors in these two compounds are given in Table 9.2.

There is a simple explanation for the large variations of acceptor binding energies between impurities substituted on the cation sublattice and those substituted on the anion sublattice. This variation is evident in the energies quoted in Table 9.2, and it is illustrated for acceptors in GaP in Fig. 9.11. When the difference between the electronegativity of the impurity, X_I, and that of the cation or anion, X_C or X_A is large, the difference between the binding energy E_B and the effective mass value E_{EM} is also large. In this limit the extra binding energy $\Delta E = E_B - E_{EM}$ behaves like a heat of formation; i.e., it is proportional to $(\Delta X)^2$, where $\Delta X = X_I - X_{C,A}$. The constant of proportionality itself is proportional to the probability that a

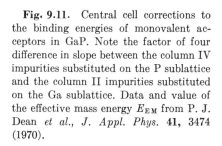

Fig. 9.11. Central cell corrections to the binding energies of monovalent acceptors in GaP. Note the factor of four difference in slope between the column IV impurities substituted on the P sublattice and the column II impurities substituted on the Ga sublattice. Data and value of the effective mass energy E_{EM} from P. J. Dean *et al.*, *J. Appl. Phys.* **41**, 3474 (1970).

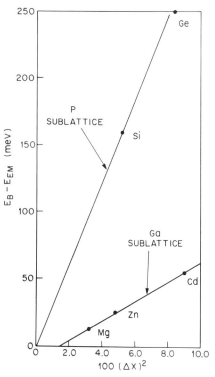

TABLE 9.2

Binding Energies of Shallow Donors and Acceptors
in GaP and CdS[a]

CdS, donors and acceptors		GaP, donors and acceptors	
Cd site	S site	Ga site	P site
In 33.8 (d)	F 35.1 (d)	Sn 65.5 (d)	O 893 (d)
Ga 33.1 (d)	Cl 32.7 (d)	Be 50 (a)	S 104.0 (d)
Li 165 (a)	Br 32.5 (d)	Mg 53.5 (a)	Se 102.0 (d)
Na 169 (a)	I 32.1 (d)	Zn 64.0 (a)	Te 89.5 (d)
	P ~1000 (a)	Cd 96.5 (a)	C 48.0 (a)
	As ~1000 (a)		Si 203 (a)
			Ge ~300 (a)

[a] The donors are labeled by (d), the acceptors by (a). The energies are in units of 10^{-3} eV, and the data were collected by Baldereschi [4].

valence electron will find itself in the atomic cell of the cation (probability P_C) or the anion (probability P_A). The explicit equation for ΔE is

$$\Delta E = E_B - E_{EM} = \mathrm{const}[P_{C,A}(X_I - X_{C,A})]^2. \qquad (9.15)$$

According to our discussion of spin–orbit energies [Eq. (7.38)], in GaP, where $f_i = 0.33$,

$$P_C/P_A = (1 - f_i)/(1 + f_i) = 0.5. \qquad (9.16)$$

Combining (9.15) and (9.16), we see that ΔE should increase four times as rapidly with $(X_I - X_A)^2$ as it does with $(X_I - X_C)^2$ for large values of ΔE. This formula is confirmed by the two slopes shown in Fig. 9.11. For acceptors in CdS, the factor $(P_C/P_A)^2 = 0.04$, indicating a very large difference in ΔE for P and As acceptors on S anion sites compared to Li and Na acceptors on Cd cation sites. This is qualitatively correct, according to the results given in Table 9.2. Notice that these differences qualitatively require ΔE to be proportional to $P_{C,A}^2$ in Eq. (9.15), not to $P_{C,A}$.

FREE AND BOUND EXCITONS

When a photon is absorbed in an electronic transition near the absorption edge of a semiconductor, an electron is promoted to the conduction band, leaving a hole behind in the valence band. The electron and hole attract each other through the Coulomb interaction, and can form a bound state of hydrogenic character. When the kinetic energies of the electron and hole are anisotropic, or when there is degeneracy at the band edge (as in the case of the valence band), then corrections to the hydrogenic model are necessary, just as for the impurity states discussed above. If the electron kinematics are described by a density-of-states mass m_e^* and the hole kinematics by m_h^*, the exciton binding energy is given by

$$E_{exc} = (\mu^*/m)(1/\bar{\epsilon}^2) \text{ Ry}, \qquad (9.17)$$

where μ^* is the reduced exciton mass,

$$1/\mu^* = 1/m_e^* + 1/m_h^* \qquad (9.18)$$

and $\bar{\epsilon}$ is given by Eq. (9.7). Some values of E_{exc} are given in Table 9.3. The hydrogenic model (9.17) is particularly useful for excitons, because there are no corrections to E_{exc} arising from "central cell" effects, inasmuch as there are no impurities to produce such central cell or electronegativity corrections.

Excitons can also be bound to neutral donors or acceptors. When this happens we have a situation involving three moving particles bound to the fixed impurity potential. In the case of donors, for example, there are two electrons and one hole bound to the positively charged impurity center. The contribution of ionic polarization of the crystal to the binding energy depends on $\bar{\epsilon}$, the average polarizability, and this factor varies depending on the disposition of the three particles. Nevertheless, theoretical calculations for donor–exciton complexes have yielded binding energies W of

TABLE 9.3

Exciton Binding Energies in Several
Semiconductors[a]

Crystal	E_b, experimental (meV)	E_b, theoretical (meV)
Ge	1.2	1.4
GaAs	3.8	3.8
InSb	—	0.6
ZnSe	19	23
CdTe	10	11.7

[a] Experimental data and theoretical values from A. Baldereschi and N. O. Lipari, *Phys. Lett.* **25,** 373 (1970). Data for GaAs from D. Binberg and W. Schairer, *Phys. Rev. Lett.* **28,** 442 (1972).

TABLE 9.4

Theoretical and Experimental Values for
the Binding Energy W of Excitons to
Donors[a]

Crystal	W, theoretical (meV)	W, experimental (meV)
InSb	0.7	—
InP	6.8	—
GaAs	6.2	6.1
GaP	22.8	—
AlSb	12.8	—
GaSb	2.7	—
CdS	34	32
CdSe	27	—
CdTe	15	14
ZnS	77	—
ZnSe	37	—
ZnTe	13	—
ZnO	70	68

[a] Values from G. Mahler and U. Schröder, *Phys. Rev. Lett.* **27,** 1358 (1971).

the exciton to the donor. These are listed in Table 9.4, together with the experimental data. Where data are available, the agreement between theory and experiment is good.

Exciton complexes can also be formed through binding to isovalent impurities. For N in GaP, this process can be visualized in stepwise fashion as an electron binding to the N atom because of the electronegativity difference between N and P. A hole is then attracted by the Coulomb potential of the electron. This is equivalent to having one exciton bound to the N isovalent impurity. Luminescence from a complex of two excitons bound to a N impurity in GaP has been observed. In this case exchange between the two electrons is essential to forming the two-exciton bound state.

DONOR–ACCEPTOR AND ISOVALENT PAIRS

Our discussion of the properties of impurities in semiconductors has emphasized so far the case of isolated impurities. However, when more than one kind of impurity is present, interactions between the impurities may lead to the observation of pairing effects. The way in which this can happen is particularly clear for monovalent donors and acceptors [9]. If the donor and acceptor lie sufficiently close to one another that there is some overlap of the donor electron wave function with the acceptor hole wave function, the electron and hole may recombine to emit a photon (luminescence). After recombination the positively charged donor and negatively charged acceptor have the Coulomb interaction energy

$$E_c = -e^2/\epsilon R, \tag{9.19}$$

where R is the donor–acceptor separation. Before the electron and hole recombined, however, the donor and acceptor were neutral centers, and there was no Coulomb interaction of the form (9.19). (This assumes that the two impurities are far enough apart so that before luminescence the screening of each center by the bound electron or hole is nearly complete). Thus the energy (9.19) is supplied to the emitted photon of frequency ω,

$$\hbar\omega = \Delta E_{cv} - E_D - E_A + e^2/\epsilon R. \tag{9.20}$$

Because the donor and acceptor are substitutional impurities, the allowed values of R in equation (9.20) are discrete. They depend on whether the two impurities lie on the same sublattice (type I) or on different sublattices (type II). Different patterns are obtained in the two cases, which are illustrated in the upper part of Fig. 9.12 (C + S in GaP, both impurities on the P sublattice, type I) and in the lower part of Fig. 9.12 (Mg on Ga sublattice and S on P sublattice, type II). The detail and richness of these luminescent spectra are remarkable. The number of shells that can be resolved seems to be limited mainly by human patience, lines having been identified as far out as the 82nd shell.

The spectra shown in Fig. 9.12 cut off at a smallest shell number of seven or eight. The reason for this is that when the donor and acceptor are closer together than a certain critical value (of order twice the Bohr radius of the donor or acceptor 1s ground state), the total energy of the system is lower for a free exciton than for one localized at the donor–acceptor pair. Alternatively, when the donor and acceptor are too close together, the attractive potential for the electron or hole becomes dipole-like and is no longer strong enough to bind both particles.

Because the donor and acceptor pair spectra are so sharp, the sum $E_D + E_A$ of donor and acceptor-binding energies can be determined very accurately from equation (9.20). Information on multipole moments of the substituted atoms can be obtained from corrections to equation (9.20).

Luminescence has also been observed from electrons and holes bound to pairs of isovalent impurities such as N in GaP. However, now there is no

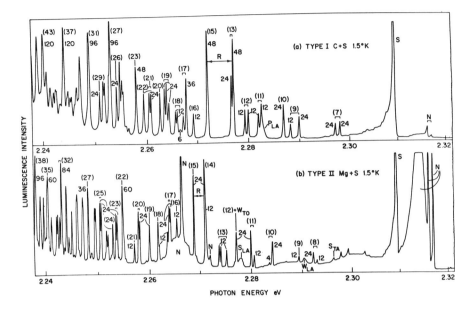

Fig. 9.12. Photoelectric recordings of luminescence spectra of donor–acceptor pairs in GaP. Most of the lines are associated with pairs indexed by their relative separation (shell number in parentheses). The intensity of lines from nearby shells is roughly proportional to the number of atoms in a given shell, shown by the numbers not in parentheses. [Data from P. J. Dean and L. Patrick, *Phys. Rev.* **2B**, 1888 (1970).]

Coulomb energy [such as (9.20)] which can be used to index the separations of different pairs. One can say that the intensity of any luminescence line involving an isovalent pair should go as the square of the impurity concentration, while luminescence from an exciton localized at an isolated isovalent impurity is proportional to the concentration of that impurity. In this way one can identify the pair lines and estimate the concentration of the isovalent impurity to within about a factor of five.

SELF-COMPENSATION

One of the most important properties of semiconductors for practical applications is that many of them can be doped with both donor and acceptor impurities. This makes it possible for one part of the crystal, where there is a preponderance of donor impurities, to contain an excess of negative electrons. This part of the crystal is called n-type. At the same time, in another part of the same crystal there may be a preponderance of acceptors and hence an excess of positive holes (p-type region). Most semiconductor devices utilize such regions of variable charge density, and hence it is important to know whether a given material can be made both n-type and p-type. Such materials are called *ambipolar*.

Elemental semiconductors (Si, Ge) and most III-V compounds, such as GaAs, are ambipolar. However, the more ionic semiconductors such as most II-VI compounds tend to be unipolar, i.e., they exhibit either n-type or p-type conductivities but not both. The mechanism which explains this behavior is *self-compensation*. Suppose, to be specific, that a halogen donor such as I replaces Te in ZnTe. Suppose further that it costs energy $\Delta H_{\rm v}({\rm Zn})$ to create a Zn vacancy in ZnTe, and that this vacancy is a divalent acceptor; i.e., it binds two holes. The binding energies $E_{\rm D}$ of the I donor and $E_{\rm A}^{1,2}$ of the Zn vacancy acceptor are, for simplicity, neglected compared to the minimum energy $\Delta E_{\rm cv}$ between the top of the valence band and the bottom of the conduction band. According to Table 7.1, this energy is about 2.2 eV.

Now suppose that a Zn vacancy is created in I-doped ZnTe. This costs energy $\Delta H_{\rm v}({\rm Zn})$. However, the donor electrons associated with two I atoms can now drop in energy from near the conduction band edge to occupy the acceptor states around the Zn vacancy. The energy of the latter is close to that of the valence band edge. Thus energy nearly equal to $2\Delta E_{\rm cv}$ is gained back. If this energy is greater than $\Delta H_{\rm v}({\rm Zn})$, then energy is gained overall by creating the vacancy and trapping the donor electrons, which are no longer available for thermal excitation as n-type current carriers.

The argument as given here neglects the binding energy of the vacancy compared to $\Delta E_{\rm cv}$, but in borderline cases (such as the II–VI compounds) this is not a good approximation [10]. If one assumes that cation vacancies (such as Zn) are acceptors and anion vacancies (such as Te) are donors, then one can reasonably correlate vacancy-binding energies with vacancy size. In CdS the smaller S anion vacancies are expected to have smaller binding energies than the larger cation vacancies. Thus the S anion vacancies compensate p-type impurities better than the Cd cation vacancies can compensate n-type impurities, and CdS is an n-type unipolar compound. The same argument applies to CdSe, ZnSe and ZnO. In the case of ZnTe the anion is larger and ZnTe is unipolar p-type. In the case of CdTe, $\Delta E_{\rm cv}$ is quite small (1.6 eV) and self-compensation is not complete: CdTe is ambipolar. Finally, ZnS is insulating, presumably because it is stable in both cubic and wurtzite structures. Thus crystals of ZnS probably contain many stacking faults with an abundance of trapping sites available for both electrons and holes.

POLYVALENT IMPURITIES

Our discussion so far has emphasized isovalent and monovalent impurities, because the solubility of polyvalent impurities in tetrahedral semiconductors is small. When the number of valence electrons of the impurity differs from that of the host atom by two or more, the excess electrons (or holes, if there is a deficiency of valence electrons) weaken the semiconductor bonds. Moreover, in the melt the impurity can associate with other atoms and use all its valence electrons for some kind of bonding. This leads to very low values of the distribution coefficients (9.1) and correspondingly small solubilities for polyvalent impurities.

As examples of polyvalent impurities consider the divalent impurities Mg and S in Si. Divalent impurities would be expected to have helium-like spectra in the same effective-mass model that predicts hydrogenic spectra for monovalent impurities. However, the ionization energy of He is nearly four times that of hydrogen. We have already seen that central cell corrections spoil the simplicity of the hydrogenic model for ground state energies of monovalent substitutional impurities, and these corrections increase in importance for deeper states. On the other hand, excited p states of substitutional divalent S impurities in Si still exhibit the effective mass pattern shown in Fig. 9.9 for monovalent impurities. To the excited p electron, the divalent impurity with one electron in a 1s ground state looks like an ionized monovalent impurity.

When Mg impurities were added to Si, they were found to behave electrically as donors. Had the Mg impurity occupied a substitutional site, it would have behaved as an acceptor. Thus one can conclude that Mg impurities enter the lattice interstitially, like Li. However, according to Fig. 9.10, Li impurities in Si have ground state binding energies very close to the hydrogenic effective mass value of about 32 meV. The ionization energies of Mg are $E_1 = 107$ meV and $E_2 = 256$ meV, compared to helium-like values of about 56 and 128 meV, respectively. The important difference between Li and Mg lies in their ion cores, the Mg $2s^2 2p^6$ outer core electrons being more numerous and less tightly bound than the Li $1s^2$ core electrons. This presumably causes a greater expansion of the Si host lattice around the Mg interstitial than it does around the Li interstitial. Although this argument predicts a larger central cell correction for interstitial Mg than for interstitial Li, it is not enough to suggest that the Mg correction should be 100 meV relative to a Li correction of order 1 meV. It is reasonable to suppose that the interstitial Mg whose levels are observed optically is associated either with a vacancy or with another interstitial Mg. Divalent S substitutional donors in Si are often associated, according to spin resonance studies, and association of polyvalent impurities with each other and with other impurities and vacancies may be common even at very low concentrations.

TRANSITION METAL IMPURITIES

Much of our knowledge of electrical behavior and lattice positions of transition metal impurities like Fe in semiconductors such as Si has been obtained through electron spin resonance studies [11]. The transition metal atom usually enters the lattice interstitially, at very low concentration, ($\sim 10^{15}$/cc), and often associates with other transition metal atoms, to form pairs or even clusters of four atoms. A few examples of substitutional transition metal impurities are known. The transition metal impurities which have been most studied in Si are Fe, Cr, Mn, and V. All transition metal impurities give rise to deep-lying electronic levels close to the middle of the energy gap. Most of them can be charged $+$, $-$, or 0 depending on the shallow impurities present in the sample. In other words, the transition metal impurities trap electrons or holes associated with shallow donors or acceptors, thereby reducing the electrical activity of the crystal. In this respect Cu, Ag, and Au behave as transition metal impurities. This is apparent from the energy levels shown in Figs. 9.3 and 9.4.

SUMMARY

Semiconductor technology has produced the most perfect and pure materials known. This has made possible controlled studies of impurities and their properties, with attention focused primarily on readily soluble monovalent substitutional donors and acceptors. The bound states of isolated impurities as well as impurity pairs have been analyzed and explained quantitatively in the simpler cases.

REFERENCES

1. R. A. Laudise, "The Growth of Single Crystals," p. 195. Prentice-Hall, Englewood Cliffs, New Jersey, 1970.
2. C. D. Thurmond, *in* "Properties of Elemental and Compound Semiconductors" (H. C. Gatos, ed.). Wiley (Interscience), New York, 1960.
3. H. Reiss and C. S. Fuller, *in* "Semiconductors" (N. B. Hannay, ed.). Van Nostrand Reinhold, Princeton, New Jersey, 1959.
4. A. Baldereschi and J. J. Hopfield, *Phys. Rev. Lett.* **28,** 171 (1972).
5. W. Kohn, *Solid State Phys.* **5** (1957).

6. S. D. Mahanti and C. M. Varma, *Phys. Rev. Lett.* **25,** 1115 (1970).
7. G. Dresselhaus, A. F. Kip, and C. Kittel, *Phys. Rev.* **98,** 368 (1955).
8. R. A. Faulkner, *Phys. Rev.* **184,** 713 (1969); A. Baldereschi and J. J. Hopfield, *Phys. Rev. Lett.* **28,** 171 (1972).
9. J. J. Hopfield, D. G. Thomas, and M. Gershenzon, *Phys. Rev. Lett.* **10,** 162 (1963); J. J. Hopfield, *Proc. Int. Conf. Phys. Semiconduct.* p. 725. Dunod, Paris, 1964.
10. G. Mandel, F. F. Morehead and P. R. Wagner, *Phys. Rev.* **136,** A826 (1964).
11. G. W. Ludwig and H. H. Woodbury, *Solid State Phys.* **13,** 223 (1962).

10

Barriers, Junctions, and Devices

The extraordinary growth of interest in semiconductors is attributable to electronic devices such as the transistor, invented by J. Bardeen, W. Brattain, W. Shockley, and coworkers in the late 1940's and following years. More than 250 books alone have been written on this subject, together with thousands of review articles and original papers. The aim of this chapter is to provide the reader with a concise qualitative description of these devices and of the physical and chemical principles behind their operation.

Explicit formulas for the current response of specific devices, derived on the basis of conventional transport theory, can be found in many books [1]. Another important question, however, is concerned with the structural origin of the specific material properties of semiconductors which make them so suitable for device purposes. This question has often been neglected, with the result that many readers have been left with the feeling that transistor behavior might be obtainable in almost any insulator, even a glass. We will discuss here device characteristics with special emphasis not on charge flow, but rather on those properties of the host material itself which are absolutely essential to practical success.

FERMI LEVELS

When a semiconductor is doped with donor and acceptor impurities, and the concentration of each kind of impurity varies from one region of the sample to another, the electronic energy levels vary as well. In the absence of an external electric field (applied potential), one quantity remains constant, the Fermi level or chemical potential μ of an electron. If this quantity were to vary, electrons would flow from regions where the Fermi level was higher to regions where it was lower until the value of μ was actually constant. Thus determination of μ for a given concentration of impurities in one region of the sample fixes the position of the electronic energy levels and carrier concentrations in that region relative to other regions.

Statistical relations between concentrations of electrons and holes and the concentrations of donor and acceptor impurities are conveniently expressed in terms of a temperature-dependent electron density parameter

$$n_{0e} = 2(2\pi m_e^* kT/h^2)^{3/2}, \tag{10.1}$$

where m_e^* is the average effective mass for electrons and a similar quantity for holes,

$$n_{0h} = 2(2\pi m_h^* kT/h^2)^{3/2}. \tag{10.2}$$

There is one quantity which is independent of μ. This is the product of the electron density n and the hole density p, which is governed by the law of mass action:

$$np = n_{0e}n_{0h} \exp(-\Delta E_{cv}/kT). \tag{10.3}$$

At 300°K the value of np is about $2 \cdot 10^{20}$ cm^{-6} for Si and $6 \cdot 10^{26}$ cm^{-6} for Ge. In the absence of impurities the crystals are intrinsic and $n = p$, so that the carrier concentrations in intrinsic Si and Ge at 300°K are $1.4 \cdot 10^{10}$ cm^{-3} and $2.5 \cdot 10^{13}$ cm^{-3}, respectively. In practice the concentrations of shallow impurities exceed these figures in almost all cases, so that almost all of the carriers are obtained from ionized impurities.

In the intrinsic limit the numbers of electrons and holes are equal, and their Fermi factors are nearly equal. The Fermi factor for an electron of energy ϵ is

$$f_e(\epsilon) = 1/\{1 + \exp[(\epsilon - \mu)/kT]\} \cong \exp[(\mu - \epsilon)/kT], \tag{10.4}$$

when $(\mu - \epsilon) \gg kT$. The Fermi factor for a hole is

$$f_h(\epsilon) = 1 - f_e(\epsilon) \cong \exp[(\epsilon - \mu)/kT]. \tag{10.5}$$

But $\epsilon_e - \epsilon_h \cong \Delta E_{cv}$, so from (10.4) and (10.5) one finds that μ lies approximately halfway between the top of the valence band and the bottom of the conduction band.

In the more realistic cases where the value of μ is determined by the impurity concentration, we may consider three possible situations: (a) impurities of only one kind, donors or acceptors but not both; (b) impurities of both kinds, but one of them much more concentrated than another; and (c) equal numbers of donors and acceptors.

In case (a), if the binding energy E_0 of the impurity is comparable to kT and is much less than ΔE_{cv}, then some of the impurities will be ionized, and some will be occupied. With the Fermi factor (10.4) changing rapidly only for $\mu \cong \epsilon = E_0$, this places the Fermi level at $\mu \cong E_0$. Thus if the impurities are donors μ falls near the bottom of the conduction band; if they are acceptors μ falls near the top of the valence band. This case is called uncompensated, meaning impurities of only one kind.

In case (b), there are majority and minority impurities (partially compensated). Suppose the majority impurities are donors and the minority impurities are acceptors. Electrons leave donor energy levels near the bottom of the conduction band to occupy acceptor energy levels near the top of the valence band until the latter are completely full. This leaves the majority donor levels partially occupied, so that μ still falls near E_0, but now it is a few kT lower than it was in case (a).

In case (c), we have virtually total compensation when the numbers of donors and acceptors differ by an amount comparable to the intrinsic concentrations quoted above. Because these concentrations are so small, total compensation is seldom achieved in a homogeneous sample. If it were, the Fermi level μ would again fall at about the middle of the energy gap.

The problem of determining the Fermi level in general is a tedious one, which can be solved graphically [1]. However, the foregoing discussion shows that in practice μ lies near the bottom of the conduction band in n-type material, and near the top of the valence band in p-type material.

BAND BENDING

Suppose a semiconductor sample is formed in the shape of a rod, with one end doped with donors (n-type) and the other end doped with acceptors (p-type). Because the sample is not homogeneous, the usual energy-band diagram of E as a function of crystal momentum $\hbar\mathbf{k}$ is not appropriate. Instead plots are made of the energy E_c near the bottom of the conduction band and E_v near the top of the valence band as a function of coordinate z along the rod. In practice the impurity gradients and space charge gradients are such that the n-type and p-type regions are separated by distances of order hundreds or thousands of lattice constants. Then the regions near the bottom of the conduction band or top of the valence band are broadened, according to the uncertainty principle, by amounts small compared to either kT or the impurity binding energies, and $E_c(z)$ and $E_v(z)$ are well defined quantities. We say that the energy bands bend because of the n-type and p-type doping.

A p–n junction can therefore be illustrated by the energy band-edge diagram shown in Fig. 10.1. On the left end of the sample donors are the majority impurity; on the right end acceptors are the majority impurity. The transition region between the n-type and p-type regions is typically a few microns wide. On the left end almost all the carriers are negative electrons, while the majority of the carriers on the right end are positive holes. The total carrier charge density satisfies Poisson's equation, and varies gradually from negative to positive as z increases through the transition region. Thus it passes through zero in this region, which is also called the depletion region.

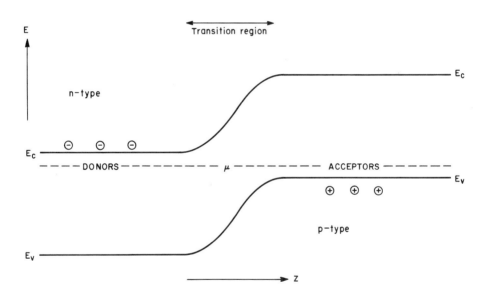

Fig. 10.1. Band-edge energies E_c (conduction band) and E_v (valence band) as a function of length z along a rod, the left end of which is n-type, the right end p-type. In the absence of an external bias the chemical potential μ is constant, and is nearly equal to the ground-state donor energy $E_c - E_D$ (or bottom of the conduction band E_c on the left end) and the ground-state acceptor energy $E_v + E_A$ (which is approximately E_v) on the right end. In the transition or depletion region the bands bend by an energy which is given approximately by $E_c - E_v \cong \Delta E_{cv} = 1.1$ eV in Si. Note that in the n-type region the majority carriers are electrons from partially ionized donors, while in the p-type region the majority carriers are holes from partially ionized acceptors.

METAL–SEMICONDUCTOR CONTACTS

A second important example of band-bending occurs at metal–semiconductor contacts. These are more complicated than p–n junctions because of the presence of an interfacial region between the metal and the semiconductor. The interfacial region may be an oxide or may contain a large concentration of impurities. If the semiconductor surface has been cleaned before deposition of the metal, or if the interface has been purified by some other method, then one speaks of an ideal metal–semiconductor contact. There are two limiting cases to consider. In the first case no free charge accumulates in the interfacial or surface region. The contact potential between the metal and the semiconductor is screened by space charge (accumulation of electrons or holes) in the semiconductor near the surface. This is the Schottky limit. In the second case the interfacial region is highly polarizable, and almost all of the contact potential between the semiconductor and the metal is neutralized at the interface. This is what happens at contacts between metals and Si or Ge, and this behavior has been explained by Bardeen in terms of *surface states* at the metal–semiconductor interface. To relate the rather ubiquitous surface states to the chemical bonds between the metal and the semiconductor, it is helpful to describe band-bending in the interfacial region in these two limiting cases.

Consider first the situation in which there are no surface states and the clean metal surface and the clean semiconductor surface are not yet in contact. This is shown in Fig. 10.2a. The work function of the metal φ_m is the difference in energy between the vacuum level ($E = 0$) and the Fermi energy or chemical potential μ in the metal. We assume that the semiconductor sample is n-type, so that in the interior of the semiconductor μ is closer to E_c than it is to E_v. The difference in energy between vacuum and the conduction band energy E_c is called the electron affinity χ_s of the semiconductor.

In Fig. 10.2a the value of μ in the metal is different from the value of μ in the semiconductor; the difference is called the contact potential. If we connect the two samples by a wire, electrons will flow from one sample to the other until the chemical potentials are equalized and the contact potential is reduced to zero. In this way an electric field is created in the gap δ between the metal and the semiconductor, because (in the case $\varphi_m > \chi$ shown in Fig. 2b) negative charge has accumulated on the surface of the metal and positive charge on the surface of the semiconductor.

Now we reduce δ to zero, bringing the metal and semiconductor into ideal contact. As shown in Fig. 2c and Fig. 2d, as δ is reduced to zero, the positive charge on the surface of the semiconductor increases in magnitude and spreads into the interior region, giving rise to band-bending in a transition

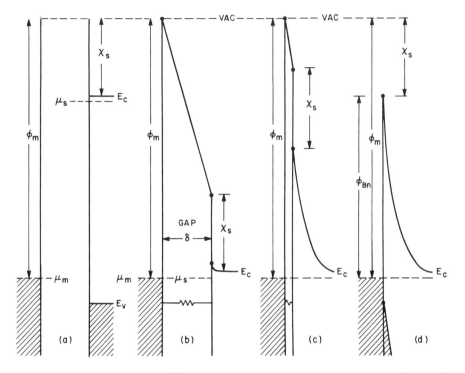

Fig. 10.2. Band-bending at a Schottky metal–semiconductor contact. In (a) the metal and semiconductor are separated, and all energies are measured relative to vacuum (VAC). In (b) the samples are joined by a wire, which makes $\mu_s = \mu_m$. In the gap of length δ between the samples there is a potential drop $\varphi_m - \chi_s$. As δ is reduced to a few microns substantial surface charge accumulates on the metal and semiconductor, and some of this spreads out into the interior of the semiconductor, causing the band-bending shown in (c). Finally, when $\delta = 0$ (ideal metal–semiconductor contact), band-bending accounts for all the barrier height $\varphi_{Bn} = \varphi_m - \chi_s$, as shown in (d).

region near the surface. This transition region is similar to that shown for the p–n junction in Fig. 10.1. Just at the interface, the maximum energy of E_c is no longer the vacuum, but is above the Fermi energy by the barrier height

$$\varphi_{Bn} = \varphi_m - \chi_s, \qquad (10.6)$$

where it is assumed that $\varphi_m > \chi$. This is Schottky's result: it says that the barrier height or work function for emitting an electron from the metal into the semiconductor is reduced from the vacuum value φ_m by the electron affinity χ_s of the semiconductor.

The expression (10.6) for the barrier height holds for an n-type insulator. For a p-type insulator χ_s must be replaced by $\chi_s + \Delta E_{cv}$, and the sign of

φ_{Bn} must be reversed (since now holes are the majority carrier). This gives

$$\varphi_{Bp} = \Delta E_{cv} + \chi_s - \varphi_m, \tag{10.7}$$

and combining (10.6) and (10.7) gives

$$\varphi_{Bn} + \varphi_{Bp} = \Delta E_{cv}, \tag{10.8}$$

a relation which is independent of φ_m. This relation is used to check the consistency of measurements of φ_{Bn} and φ_{Bp}.

Most studies of metal–insulator barriers during the 1930s utilized quite ionic insulators, for which (10.6) is valid. When barrier heights were measured for metal–Si or metal–Ge contacts, however, it was found that (10.6) was not valid. To a first approximation the barrier height φ_{Bn} is *independent* of the work function φ_m of the metal. This surprising result was explained by Bardeen in terms of surface states.

The situation envisaged by Bardeen is illustrated in Fig. 10.3. Once again electrical contact is established between the metal and the semiconductor in Fig. 10.3b. Now, however, as the gap separation δ is decreased, and the charge on the surface of semiconductor increases, all of this charge is accommodated in surface states, and there is no space-charge accumulation or band-bending in the interior of the semiconductor near the surface. As a result, φ_{Bn} is independent of φ_m and is a constant characteristic $\varphi_{Bn}(s)$ characteristic of the semiconductor s.

One can describe behavior of φ_B intermediate between that of the two limiting types through the equation

$$\varphi_{Bn} = S(s)(\varphi_m - \chi_s) + \varphi_{Bn}(s) \tag{10.9}$$

with $0 \leq S(s) \leq 1$. Broadly speaking, $S(s) \cong 0.1$ for Si and Ge, while $S(s) \cong 1.0$ for ionic crystals such as SiO_2 and Al_2O_3. The transition from small S to large S appears to take place in II–VI compounds such as CdS and CdSe.

The localization of charge in surface states of the Bardeen type requires the availability of states in the interfacial region, which would not be possible if bonds were formed between the metal and the semiconductor anions.

In Chapter 9 we saw that transition metal impurities in Si and Ge behave as deep traps which can be charged $+$, $-$, or 0 as dictated by the position of the Fermi level. Presumably the surface states which neutralize the contact potential between the metal and Si or Ge behave similarly. On the other hand, metals deposited on SiO_2 may themselves bond strongly to the oxygen anions, leaving no states available for charging at the interface.

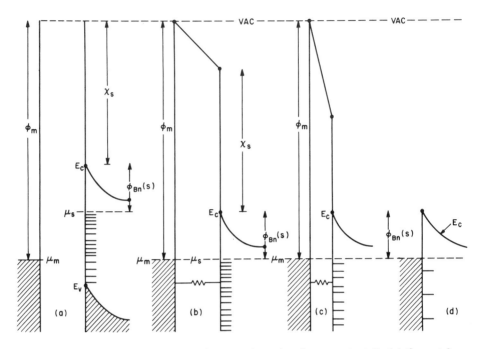

Fig. 10.3. Band bending at a Bardeen metal–semiconductor contact. In (a) the metal and semiconductor are separated, but the bands of the semiconductor are bent near the surface because electrons occupy surface states between μ_s and E_v, as indicated by the short lines. In (b) the samples have been connected by a wire, bringing μ_m and μ_s together. In (c) the gap between the samples is reduced, but the increased electric field at the surface of the semiconductor is screened from the interior by emptying some of the surface states, so that the band-bending remains almost unchanged. In (d) ideal contact has been established and more surface states have been emptied, leaving the band-bending the same and giving the barrier height $\varphi_{Bn}(s)$ independent of φ_m.

p–n JUNCTIONS

The band-bending which takes place in the transition region of a p–n junction is shown in Fig. 10.1, and again in Fig. 10.4, where the p–n junction is part of an electrical circuit with initially zero applied bias. Because the electrons are more numerous on the left side, there is an electron current from left to right which consists of the fraction of electrons incident on the barrier with enough thermal energy to climb over. If the flux of incident electrons is represented by a current J_0, the thermal fraction is given by the Boltzmann factor $\exp(-\varphi_B/kT)$, and the electron current to the p-side is given by

$$J_{\text{nr}}(0) = J_0 \exp(-\varphi_{\text{B}}/kT). \tag{10.10}$$

When the electrons have reached the p side, they recombine with the holes there, which is the reason for the subscript r in (10.10).

In the absence of an external bias, no net current flows across the junction. This is because the electron current J_{nr} from left to right is balanced by a current of electrons from right to left. This current arises from the small number of electrons always present in the p-type region because of thermal excitation or generation; these electrons can slide down the barrier from right to left, giving up their kinetic energy to the lattice in the form of heat. This current is denoted by J_{ng}, and by detailed balance

$$J_{\text{nr}}(0) + J_{\text{ng}}(0) = 0. \tag{10.11}$$

Suppose now that a positive voltage is applied to the p region and a negative voltage to the n region, thereby making $\mu_{\text{n}} = \mu_{\text{p}} + eV(V > 0)$. This is a *forward bias*, and it reduces the barrier height from φ_{B} to $\varphi_{\text{B}} - eV$, as shown in Fig. 10.5. This changes J_{nr} from (10.10) to

$$J_{\text{nr}}(V > 0) = J_0 \exp[-(\varphi - eV)/kT] = J_{\text{nr}}(0) \exp(eV/kT), \tag{10.12}$$

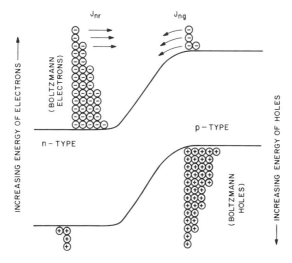

Fig. 10.4. A p–n junction with zero applied bias. The thermally activated electron recombination current $J_{\text{nr}}(0)$ is balanced by the backflow current $J_{\text{ng}}(0)$ of electrons thermally generated in the p-type region which drift back into the n-region.

while the backflow current remains unchanged:

$$J_{ng}(V > 0) = J_{ng}(0) \qquad (10.13)$$

The total current for $V > 0$ is given by

$$I(V > 0) = J_{nr}(V > 0) + J_{ng}(V > 0) = I_0[\exp(eV/kT) - 1], \quad (10.14)$$

where I_0 is a constant characteristic of the junction.

We can now reverse the voltage (*reverse bias*) and repeat the argument. With reverse bias $\mu_n < \mu_p$ and $J_{nr}(V < 0) < J_{ng}(V < 0)$, because now the barrier to be surmounted by thermal activation is greater. As a result, we find that the analysis leading to (10.14) goes through equally well for reverse bias $(V < 0)$, and that (10.14) describes the I versus V characteristics of the junction for net current flow in either direction.

For $eV/kT \ll 1$, Eq. (10.14) gives ohmic behavior,

$$I \cong I_0 eV/kT, \qquad (10.15)$$

with current linear in voltage. However, for $|eV| \gg kT$, the current increases exponentially with forward bias, and saturates with reverse bias, thus effectively passing current in the forward direction very well and in the reverse direction poorly. This is the rectifying characteristic of p–n junctions.

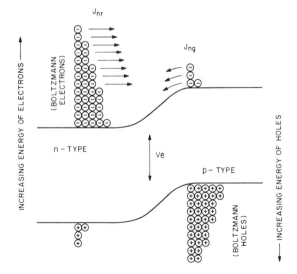

Fig. 10.5. A p–n junction under forward bias V. The reduced barrier height increases the fraction of electrons in the n-region which can drift into the p-region.

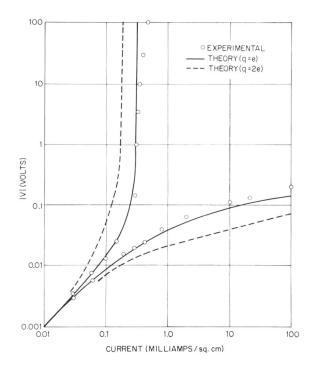

Fig. 10.6. The I vs. V characteristic of a nearly ideal p–n junction. The theoretical curves are shown for two values $q = e$ and $q = 2e$ of the electrical charge of the carriers.

The behavior of an ideal p–n junction is described by Eq. (10.14), and p–n junctions made out of Ge satisfy (10.14) quite well, as shown in Fig. 10.6. Incidentally, the good agreement between theory and experiment shows that the charge q carried by holes and electrons is $\pm e$, as we would have expected. If the charge q were, say $\pm 2e$, the dashed lines in Fig. 10.6 would be expected (more strongly rectifying).

In practice Eq. (10.14) is found to be an idealization which gives an upper limit to the rectifying ability of the p–n junction. When the carrier concentrations in the n-type and p-type regions are large (as they are in Ge at room temperature, because of its smaller value of 0.8 eV for ΔE_{cv}), then the discussion given above includes the principal contributions to I. For smaller carrier concentrations in Si and GaAs junctions ($\Delta E_{cv} = 1.1$ and 1.4 eV, respectively), recombination and generation take place to an appreciable extent in the barrier region itself. As a result, the effectiveness of the barrier in differentiating the two currents is reduced, and a modified form of (10.14) is often used to describe the I versus V characteristics of Si or GaAs junctions. This is

$$I \cong I_0[\exp(eV/nkT) - 1],\tag{10.16}$$

with $n \cong 2$.

CARRIER INJECTION AND TRAPPING

The discussion of the rectifying action of p–n junctions and metal–semiconductor contacts just given emphasizes the electron current, but the same discussion applies (with appropriate reversals of sign) to the hole current as well. In an n-type region most of the current is carried by electrons, while in a p-type region most of the current is carried by holes. Near a semiconductor surface band-bending may bring the Fermi level closer to the top of the valence band, although in the n-type bulk it was closer to the bottom of the conduction band. This situation is implicit in Fig. 10.3 (metal–semiconductor contacts), but for clarity it is repeated in Fig. 10.7.

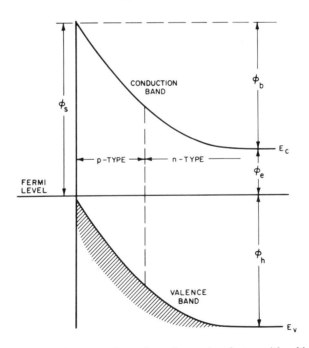

Fig. 10.7. Band-bending near the surface of a semiconductor with a high density of Bardeen–Shockley surface states. Although the material is n-type in the bulk, it is p-type near the surface, in a region called the depletion layer (from Bardeen and Brattain [2]).

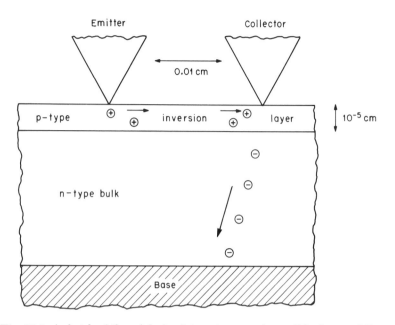

Fig. 10.8. A sketch of the original point-contact transistor of Bardeen and Brattain [2].

The p-type region near the surface is called an inversion layer. In this region the current is carried by holes, and at the interface between the inversion layer and the bulk the resistance is a maximum.

The fact that current in a semiconductor can be carried by both electrons and holes is the basis of the first transistor invented by Bardeen and Brattain [2].† They used a small block of n-type Ge with a large area base contact to draw off electron current from a point contact which was reverse-biased (high resistance). A second point contact is placed close to the first and is biased forward (low resistance). Both contacts are made to the surface inversion layer (p-type), so that the forward biased contact emits holes and the reverse-biased one collects them, as shown in Fig. 10.8.

Modulation of the collector–base electron current I_c by the emitter–collector hole current I_e apparently takes place through modification of barrier rectification at the collector contact. Presumably the holes reduce

---

† This is the classic transistor paper. In addition to containing simple theory and a practical description of the early point contact experiments, it also includes a history of the early efforts to achieve amplification in semiconductor devices. See also W. H. Brattain, *Phys. Teacher* March (1968).

the barrier height below its steady-state value, thereby increasing the electron current. Because the holes have been emitted across a low-resistance barrier, and are collected at a high-resistance one, the geometry is favorable to voltage amplification. However, the original Bardeen-Brattain experiments also showed current amplification, with the factor

$$\alpha = -(\partial I_c / \partial I_e) \tag{10.17}$$

being of order 2. This showed that the balance between recombination and regeneration at the barrier is easily disturbed by injection of minority carriers.

Most transistor action takes place through combining a low resistance input with a high resistance output current, hence the name *transfer of resistor*. In order for such action to take place, the electron or hole current must not decay too much. In the Bardeen–Brattain experiment this was accomplished by placing the emitter and collector about 0.01 cm apart and utilizing n-type Ge, where the electrons have high mobility because of their small effective masses. With improved control over the purity of materials, current decay is a less serious problem, and now Si is the primary transistor material, for reasons discussed in a later section.

The primary mechanism for decay of electron currents in semiconductors is through trapping. With the Bardeen–Brattain geometry pulsed transit experiments have been done that exhibit diffusive drift of carriers from the emitter to the collector, with a delay time consistent with other measurements of carrier mobility. The experiments also exhibit attenuation of the current through trapping.

The details of the trapping process are complex, but are likely to involve deep impurity states which are initially neutral (such as isovalent impurities or transition metal impurities). The deep state captures a hole (for example), becoming positively charged. The positive charge attracts an electron, which because of its localization near the hole recombines to emit some light. Most of the ΔE_{cv} energy which the electron and hole possessed initially, however, is emitted nonradiatively as lattice energy. Because the energy of the lattice phonons is small ($\lesssim 0.05$ eV, compared to $\Delta E_{cv} \sim 1$ eV) nonradiative emission involves a cascade process. The cascade, in turn, is made more probable by the availability of a series of levels in the forbidden gap which eventually lead to the ground-state of the trap. This means that efficient trapping centers do not consist of isolated deep impurity centers, but rather consist of aggregates of such impurities condensed or localized near dislocations or grain boundaries. Indeed, in the Bardeen–Brattain geometry most of the decay of the injected current takes place near the surface, where there is probably also a higher concentration of deep impurities and dislocations.

JUNCTION TRANSISTORS

The point-contact configuration employed by Bardeen and Brattain is complicated because of its three-dimensional geometry and inefficient because of the spreading resistance of the contacts. Shockley predicted that transistor action could be achieved by properly doping and biasing a series of opposing p–n junctions, such as the p–n–p configuration shown in Fig. 10.9. In this geometry the current flow is one-dimensional. The strongly doped p^{2+} region functions as an emitter and it contributes most of the current across the first p–n junction. The intermediate n^+ region is sufficiently thin that the holes injected from the left p–n junction diffuse across to the right n–p junction before recombining. By adjusting the voltages applied to the emitter, base, and collector, the first p–n junction is biased forward and the second p–n junction biased backward. As a result the voltage drop between emitter and base is small compared to that between collector and base, and large voltage amplification results.

The circuit equations for triode p–n–p or n–p–n junctions become more complex when current loss between junctions is included. More complicated junction geometries (such as p–n–p–n) can be used in three- or four-terminal configurations as switches (semiconductor-controlled rectifiers). These applications [1] all rely on the basic p–n junction and their success depends essentially on achieving uniform doping over prescribed and often very small dimensions.

TUNNEL DIODES

The barrier currents discussed so far have all been thermally activated and involve a Boltzmann factor $\exp(-\varphi_B/kT)$, as in Eq. (10.10). A second

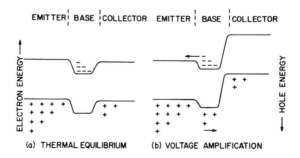

Fig. 10.9. Shockley's junction transistor. For the junction to function effectively it is necessary that the intermediate n+ region be very thin, of order a few microns or less in thickness.

way to transmit carriers through a barrier is by quantum-mechanical tunneling, a mechanism which is essentially independent of temperature. At room temperature the tunneling current exceeds the thermally-activated current in a p–n junction only when the transition region of the junction is very thin, typically 100 Å or less. This in turn requires high concentrations of impurities (n^{2+} and p^{2+}) on both sides of the junction. Typical impurity concentrations in tunnel diodes are 10^{19}–10^{20} shallow impurities/cm^3.

The I vs. V characteristic of an ideal tunnel diode is shown in Fig. 10.10. The various points correspond to the opposing energy-band configurations shown in Fig. 10.11. Because of the high doping levels, on each side of the junction the Fermi energy lies above the conduction band edge (n-type) or below the valence band edge (p-type) by several kT or more. When this is the case, the electrons on the n-type side and the holes on the p-type side are said to be degenerate.

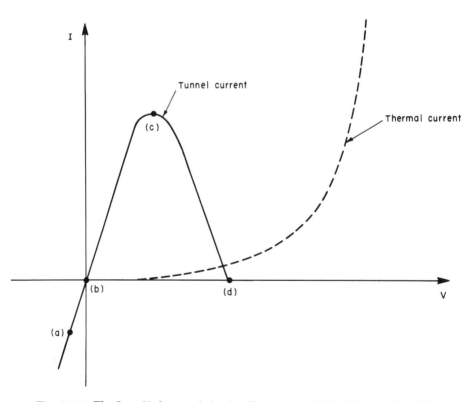

Fig. 10.10. The I vs. V characteristic of an Esaki tunnel diode. The negative differential resistance region is explained by comparing the points on the figure with the energy-band configurations shown in Fig. 10.11.

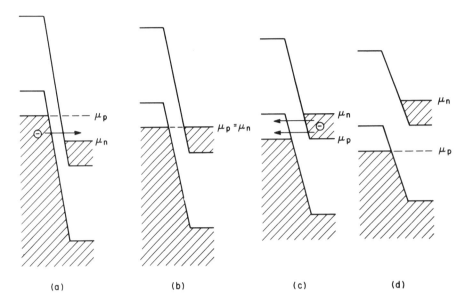

Fig. 10.11. In tunnel diodes the tunnel current is controlled by the availability of occupied initial and unoccupied final states at the same energy, allowance being made for the effect of the applied bias on the relative energies of carriers on the two sides of the high-resistance barrier.

The remarkable feature of the I vs. V curve shown in Fig. 10.10 is that it shows a large region of negative differential resistance, where an increase in voltage leads to a decrease in current. Because of this negative resistance region, tunnel diodes can be used as amplification, switching, and memory devices [1]. Because the transition regions are so thin, the time constants of tunnel diodes are much faster than those of thermally activated p–n junctions, where carriers must diffuse across the junction. Tunneling currents also exhibit low-noise characteristics because of faster time constants. The promising device possibilities of tunnel diodes were first appreciated [1] by L. Esaki in 1958.

Actual I vs. V curves for tunnel diodes made out of Ge, GaAs, and GaSb are shown in Fig. 10.12. The most important characteristic of these curves is the current peak-to-valley ratio I_p/I_v, where I_p is the peak current just before the negative differential resistance region, and I_v is the low current at the end of this region. The observed I vs. V curves shown in Fig. 10.12 differ from the idealized curve shown in Fig. 10.10 because of excess current contributions to I_v. One must pay a price for the thinness of the tunneling barrier, and this price lies in the high concentration of impurities on either side of the barrier, as well as the unavoidable sensitivity of the tunneling

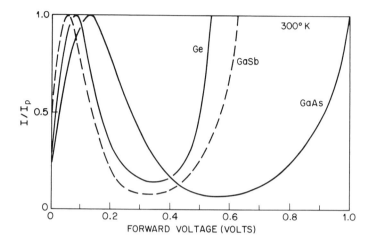

Fig. 10.12. Comparison of I vs. V characteristics of tunnel diodes fabricated from Ge, GaAs and GaSb, after Sze [1].

transmission coefficient to the apparent thickness of the barrier. The first factor smears out the true tunneling current because it leads to the formation of impurity bands in the n-type and p-type regions. These cause tailing of the valence and conduction-band edges into the energy gap, thereby degrading the definition of the edge of the tunneling current at large V (Figs. 10.10d and 10.11d). The second factor, the apparent thickness of the tunneling barrier, can be modified by clusters of donors and/or acceptors in the barrier region itself. This causes some of the carriers to tunnel not through the barrier as a whole, but partly through the barrier and partly through these impurity states, which if clustered may act as deep impurity centers.

AVALANCHE DIODES

The I vs. V curves of the tunnel diode exhibit negative resistance even at $\omega = 0$. However, in many applications operation at microwave frequencies is the practical aim, and at high frequencies differential negative resistance can be achieved by delaying an amplified current so that it reaches the electrodes out of phase with the input voltage. This is the idea behind avalanche diodes, which are useful as sources of microwave power.

Under the influence of a weak electric field ϵ carriers are accelerated until their average velocity $\bar{v}$ is equal to the drift velocity v_d. The drift velocity is proportional to ϵ according to

$$v_d = \mu\epsilon, \qquad (10.18)$$

where μ is the carrier mobility. The value of μ is determined by the scattering mechanism that limits $\bar{v}$ to v_d. At low temperatures and low values of ϵ, the predominant scattering mechanism in nearly pure crystals is one of emission of acoustic (low-energy) phonons.

Now let us suppose that the electric field is gradually increased. At first new scattering mechanisms (such as emission of higher-energy optic phonons) come into play, and these tend to reduce v_d below the value expected from (10.18). Scattering by optic phonons causes v_d to saturate near 10^7 cm/sec, and this causes the current to tend towards a saturation value. However, if there are impurities present, then the carriers may gain enough energy to liberate additional carriers through impact ionization. In this way the number of carriers is multiplied, and if this contribution to the current exceeds the saturation tendencies from other scattering mechanisms, electrical breakdown or carrier avalanching will result.

Ionization rates at fields near breakdown ($\sim 3.10^5$ V/cm) are typically of order 10^4/impurity-cm. When the impurities are shallow, differences in ionization rates of donors and acceptors of as much as a factor of ten have been observed in Si, but in Ge, GaAs, and GaP the two ionization rates appear to be much the same.

The geometry of an avalanching diode involves a p$^+$–n junction followed by an intrinsic (high-resistance) region. The transition region between the highly doped p$^+$ region and the n region is one of very high electric field ϵ, and it is here in a region of width W_1 that the avalanching takes place. The carriers so created must traverse the high-resistance region of width W ($\sim$several microns) before reaching the collector. Actual diodes utilize junction widths W_1 and transit widths W such that $W_1/W \sim 0.2$.

Avalanche diodes have been given the amusing acronym IMPATT, for IMPact ionization Avalanche Transist Time. The optimal oscillation frequency is of order the transit time of the intrinsic region W. With v_d near 10^7 cm/sec and W near $3 \cdot 10^{-4}$ cm, oscillation frequencies near $3 \cdot 10^{10}$/sec are possible and have been achieved with IMPATT diodes [1].

WHY Si?

Early transistor devices were made from Ge, partly because conduction-electron effective masses are smaller there than in many other semiconductors, making the mobility in (10.18) higher and reducing the rate of current decay through trapping and recombination. With improved sample purities more favorable doping geometries and better metal–semiconductor contacts, the preferred material has become Si. There are a number of reasons for the dominance of Si in transistor technology, and almost all of these are explicable in terms of microscopic material properties.

To place the discussion on a quantitative basis it is convenient to introduce various figures of merit to compare different materials [3]. The first proposed figure of merit does involve the electron and hole mobilities μ_n and μ_p as well as the static dielectric constant ϵ_s. It is

$$Q_1 = \mu_n \mu_p / \epsilon_s^{1/2}. \tag{10.19}$$

Here both electron and hole mobilities enter symmetrically; this reflects the fact that charge neutrality requires that charge flow of carriers of one type implies charge flow of carriers of the other type as well. The dielectric constant ϵ_s enters because the flow of carriers is controlled by potential drops, and the amount of charge increases with increasing dielectric constant.

Refinements of semiconductor devices have led to the demand for higher frequencies and hence smaller critical dimensions. The ways in which these smaller dimensions have been achieved are summarized in the next section. Dimensions of order a few microns or less together with improved sample purities have made the low-field mobility figure of merit Q_1 less significant compared to other figures of merit. Operating voltages must be substantially larger than kT/e, where e is the electronic charge, in order for the rectifying properties of barriers and junctions to become effective. When these voltages are combined with small sample dimensions one approaches breakdown or avalanche conditions even if these are not desired. Moreover, in the avalanche diodes themselves high power levels mean large Joule heating. If the heating becomes too great not only will carriers be impact-ionized from impurities, they will also be created by excitation of electron–hole pairs with the gap energy ΔE_{cv}. This causes avalanching throughout the entire sample and complete breakdown of semiconducting properties, and ultimately the sample melts. The connection between maximum voltage V_m, cutoff operating frequency ω_c, breakdown field E_B and limiting carrier velocity v_L involves the figure of merit

$$Q_2 = V_m \omega_c \lesssim E_B v_L. \tag{10.20}$$

The values of $E_B v_L$ in Ge and Si are 1 and 3, respectively (10^{11} V/sec). In actual devices $V_m \omega_c$ approaches $E_B v_L$ to within a factor of 3; the limit is not practically attainable because $E < E_B$ and $v < v_L$ over parts of the charge-carrier path.

A practical point, also related to the basic properties of the material, is that silicon reacts with oxygen to form a chemically stable layer of composition intermediate between SiO and SiO$_2$. This layer protects against the environment and is an excellent electrical insulator. The silicon oxide layer packages the device, because it is nonreactive and nearly impervious to penetration by most electrically active impurities. The oxide that forms on Ge is reactive and unstable, and this is also the case with most other semiconductors. In the production of integrated circuits on Si wafers, silicon oxide can be used as a mask to protect parts of the sample from contamination while other parts of the sample are being doped.

The third and final factor that favors Si over Ge concerns the problem of heat dissipation, which as mentioned above is essential in high-power devices. The limiting factor for heat dissipation is the thermal conductivity, and the thermal conductivity of Si is three times that of Ge. The factors discussed here are compared for Si, Ge, and SiC in Table 10.1.

A survey of this table would appear to suggest that SiC is a promising material for high-power semiconductor devices. However, this is an illusion. The entire discussion is predicated on the assumption that good crystals are available at the outset. The compound SiC, however, although tetrahedrally coordinated, is uniaxial and forms in a variety of hexagonal structures. Actual samples are polycrystalline and are found with a very high density of stacking faults. In Chapter 9 while discussing the phenomenon of self-compensation we saw that it is difficult to make ZnS either n-type or p-type, because of the large concentration of stacking faults which trap free carriers. The same is true of SiC, and this makes it difficult to achieve transistor action in this material. Ultimately there is one figure of merit which overrides all others, and that is the structural uniformity of the starting material.

TABLE 10.1

Comparison of the Properties of Ge, Si, and SiC

Property	Ge	Si	SiC
Mobility (cm^2/V-sec)	4000	2000	250
Dielectric constant, K	16	12	7
Limiting velocity, ν_L (cm/sec)	6×10^6	10^7	10^7
Breakdown field, E_B (V/cm)	10^5	2×10^5	5×10^6
$E_B\nu_L/2\pi$ (V/cm)	10^{11}	3×10^{11}	8×10^{12}
Thermal conductivity, λ (W/cm-°K)	0.5	1.5	5
Forms SiO$_2$	No	Yes	No
Band gap (eV)	0.7	1.2	3.0

MICROWAVE EVOLUTION

One of the geometries which is used today for the junction transistor is shown in Fig. 10.13. In contrast to the early designs, such as that shown schematically in Fig. 10.9, modern designs emphasize a planar geometry in which the base and the emitter are made as small and placed as closely together as possible. This reduces carrier transit times and phase lags and makes operation at microwave frequencies possible. For the geometry shown in Fig. 10.13, the critical dimensions are the width S of the emitter stripe and the minimum thickness W_B of the base. Much effort has been devoted to reducing these dimensions to of order 1 μm or less. This effort is summarized in Fig. 10.14.

New techniques have been developed to make feasible the dimensional reduction, and some of these are indicated in the upper part of Fig. 10.14. The first four of these are shown schematically in Fig. 10.15. In each case a series of chemical treatments is applied to a bulk sample of n-type Si, either to remove some of the material through etching, to cover it with silicon oxide, or to dope it with impurities (usually acceptors, as the bulk is already n-type). The progressive refinement of technique is designed to reduce junction dimensions, as shown in Fig. 10.14. These first four steps cover the period 1948–1956.

Five more steps have been added in the period 1960–1970. These steps involve modifications of the bulk sample itself. In order to reduce the series resistance of the base, the latter is prepared in two stages. A heavily doped, low resistivity n+ substrate supports a thin layer of n material in contact

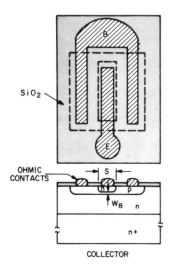

Fig. 10.13. The stripe-base geometry of some microwave transistors. The base is labeled B, the emitter E, the base depth is W_B, and the stripe width is S.

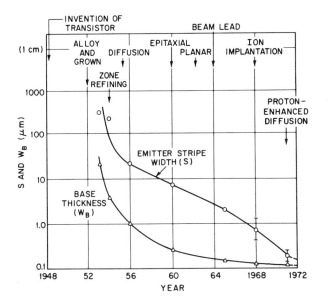

Fig. 10.14. Evolution of the two critical dimensions S and W_B shown in Fig. 10.13, as a function of time and the introduction of new materials techniques. Data up to 1968 from Sze [1], later data from H. F. Cooke, *Proc. IEEE* **59**, 1163 (1971).

with the junction. The thin layer of n-material is actually grown as a separate crystalline film, using the substrate as a master form, and the process is called epitaxy. The phrase planar technology is primarily associated, as the name implies, with preparation steps confined to a flat surface, and it makes abundant use of oxidation as a process for masking various parts of the sample before and after doping stages.

While metal–semiconductor contacts can be utilized as rectifiers, in forming integrated circuits from transistors one wants to establish nearly ohmic metal contacts that are not subject to degradation by oxidation, for example. This is accomplished in the beam lead method, where as the name implies, the leads are also used to given structural support to the silicon chip upon which rests the family of transistors composing the integrated circuit. The metal chosen for the beam lead is a transition metal such as platinum or tungsten. At temperatures below the melting point of either the metal or the silicon substrate, a solid–solid reaction takes place to form a compound transition metal silicide which penetrates several hundred angstroms below the initial surface of the silicon. The compound is itself a metal, and the contact between it and the silicon is nearly ohmic (almost nonrectifying). Moreover, the metal–semiconductor interface is now safely protected from oxidation by the external environment. Finally,

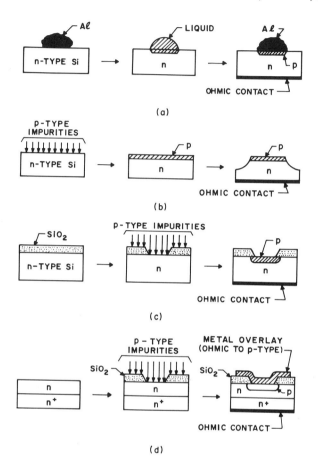

Fig. 10.15. Evolution of diffusive doping techniques, 1950–1956: (a) alloyed junction, (b) diffused mesa junction, (c) diffused planar junction, and (d) diffused planar junction on epitaxial substrate. See also the upper part of Fig. 14 (from Sze [1]).

the strong bond between the compound and the sample adds mechanical stability to the entire system.

One of the inherent limitations imposed through diffusive doping is that the impurity distribution cannot be made to resemble a step function; rather it is blurred, with Gaussian edges. Ion implantation adds the impurities to the sample by accelerating them to high energies as ions, and using these ions to bombard the sample. The great advantage of implantation is that after penetrating the surface a vertical stream of bombarding ions loses its energy at a steady rate and eventually comes to rest at a certain (nearly constant) distance below the surface, the distance being determined by the initial energy. This profile is excellent for junction preparation. The disadvantage of the method is that the bombarding ions cause severe damage to the crystal structure itself near the surface. This damage can be removed by annealing, but this in turn causes diffusion of the implanted ions, thereby blurring the profile with diffusive doping. Moreover, the implanted ions are not in thermal equilibrium, so many of them do not occupy substitutional sites and are therefore not electrically active: instead that fraction, together with the damage caused by bombardment, can act as traps which inhibit transistor action.

Fortunately it it possible to retain most of the advantages of ion-implantation, in particular the sharp definition of the emitter width S, without most of the damage-related disadvantages. Note that the ion-bombardment process uses the ions to perform two tasks: to open a path through the lattice structure, and to produce electrically active impurities after coming to rest. These two functions can be performed more success-fully by different agents. This is the idea behind proton-enhanced diffusion. By proton bombardment vacancies are created near the surface. The doping impurity is added to the surface of the crystal, and at relatively low tem-peratures the impurities diffuse through the surface region (step-function profile) where vacancies make diffusion easy. After coming to rest the protons are not electrically active, and may diffuse out of the bulk Si, possibly being trapped in the oxide. In any case the doping profile that results is an image of the original vacancy-rich proton-bombarded region. In this way both emitter widths and base depths can be defined with a resolution of order 0.1 μm which is comparable to the best results that can be achieved by, e.g., low-temperature epitaxy for base depths.

This short summary does not do justice to the many technological refinements of transistor manufacture, but it does acquaint the reader with some of them. The development shown in Fig. 10.14 probably represents the most sophisticated evolution of physical and chemical techniques known in materials science.

LUMINESCENCE

For transistor action recombination of electrons and holes plays a vital role, as we saw in discussing the current balance of a p–n junction in equilibrium. If the electron–hole annihilation takes place outside the transition or barrier region junction action will take place, but if too much recombination takes place in the active regions, transistor behavior will suffer. Thus for transistor purposes trapping and recombination are avoided as much as possible. On the other hand, if the recombination of electrons and holes can be made to take place with high efficiency, then semiconductors may be used to convert electrical energy into light. This is actually done with electroluminescent diodes.

While Si and Ge are the materials most used for transistors, their small band gaps ΔE_{cv} (below the visible portion of the spectrum) make them unattractive as electroluminescent materials. Band-gap radiation from GaAs and GaP falls near and in the visible region of the spectrum, and so is potentially useful for data display [4]. Of the two materials, GaAs has a direct band-gap of 1.4 eV (see Fig. 5.13) while GaP has an indirect band-gap of 2.26 eV (conduction-band edge similar to Si; see Table 5.2). Efficient luminescence, as we shall see, involves impurities and so takes place at energies less than ΔE_{cv}. Depending on the impurities used, GaP luminesces in the red near 1.78 eV with a quantum efficiency of 2%, or in the green near 2.23 eV with an efficiency of only 0.02%. To the human eye the apparent brightness is about the same, so detector efficiency as well as quantum efficiency is important. For this reason GaP is a more promising electroluminescent material than GaAs.

The central problem in using GaP as an electroluminescent material is that the intrinsic recombination efficiency in the absence of impurities is very low. The reason for this is that in an indirect gap material the recombination of an electron and hole requires the emission or absorption of a lattice phonon (conservation of crystal momentum). This three-particle process is much less likely than the two-particle electron–hole recombination which is allowed in a direct-gap material like GaAs. Both Ge and Si are indirect-gap materials, and this reduces their recombination rates as well. In fact, had Ge been a direct-gap material, Bardeen and Brattain probably would not have discovered point-contact transistor action in it.

The central impurity mechanism for increasing the quantum efficiency of electroluminescent diodes is the one that we have described microscopically in Chapter 9 as donor–acceptor pairs. For example, in GaP well-separated C and S pairs luminesce in the green, while nearest-neighbor Cd and O pairs luminesce in the red. The C, S, and Cd impurity states are shallow, but the O state is deep (Table 9.2). It may seem paradoxical that the C and S shallow states, which are larger in diameter and hence more likely to overlap, should radiate less efficiently than Cd and O nearest-neighbor pairs. However, the Cd and O impurities are so different from the host Ga and P atoms that there is a large probability of Cd and O association as nearest-neighbor pairs on the Ga and P sublattices, respectively. For C to be an acceptor and S to be a donor, however, they must both reside on the P sublattice and hence cannot be nearest neighbors. If they are, they are both donors and do not act to increase the efficiency of electron–hole recombination.

JUNCTION LASERS

The basic idea behind the junction laser is so simple that its realization was achieved independently and almost simultaneously by three different experimental groups. Just as the transistor performs electronic operations in the solid that vacuum tubes achieve with gaseous ions, so the junction laser replaces the gas laser. The gas laser requires an inversion of occupation of atomic energy levels in order to achieve *light amplification by stimulated emission of radiation*, and this inversion requires optical pumping. The idea behind the junction laser is much simpler: in the transition region of a p–n junction there is an abundant supply of injected electrons and holes, and this supply can be readily replenished by forward-biasing the junction. All that is needed, therefore, is to stimulate the recombination with incident light of energy near that of ΔE_{cv}, and lasing action should result.

The most obvious material for junction or injection lasers is GaAs, because it is a direct-gap semiconductor and because it is easy to dope to form n-type and p-type regions and narrow transition regions. With such junctions lasing action is obtained with a bias of 1.5 to 2 V.

Early GaAs diode lasers, however, did have one major drawback. The efficiency of stimulated emission of radiation increases with increasing light intensity. The early diode lasers required a high current density to operate, and in the junction region this produces great heating. It is just the problem of heating that makes silicon preferable to germanium for high-power transistors. But GaAs is very similar to Ge in terms of thermal conductivity and melting point, because Ga and As are adjacent to Ge in the periodic table (Fig. 2.1). This meant that GaAs junction lasers could not be operated continuously, but only with very short pulses (of order one microsecond at room temperature).

The heating problem can be overcome if some way can be found to utilize the intrinsic efficiency of electron-hole recombination and confine the process to a narrow transition region, thereby making possible laser action at lower current levels. From the examples of electroluminescence just discussed, we see that this is a materials problem which can be solved by suitable alloying and doping in the neighborhood of the junction. The solution is the double heterostructure junction laser [5] shown in Fig. 10.16.

This junction utilizes alternating layers of GaAs–$Al_xGa_{1-x}As$, with x near 0.2. The $Al_xGa_{1-x}As$ pseudobinary alloy has a larger energy gap than GaAs, because E_0 in AlAs is near 3 eV (Table 5.1) and because E_0 is nearly a linear function of x (Chapter 8). As shown in Fig. 10.16, once electrons and holes have been injected into the center slice of GaAs, the adjacent sheets of $Al_xGa_{1-x}As$ act to contain the injected carriers and to prevent their

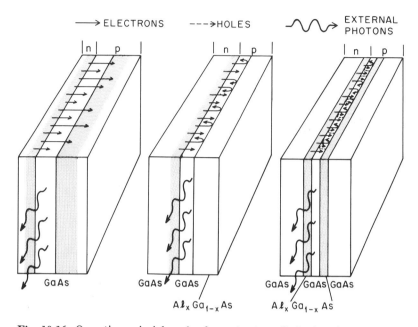

Fig. 10.16. Operating principles of a homostructure GaAs junction laser (left), a single-heterostructure laser (middle), and a double-heterostructure laser (right) are compared. In the homostructure geometry electrons injected from the n region to the p region travel through a broad region before decaying radiatively by recombining with a hole (with emission of light toward the reader) or nonradiatively (Joule-heating). In the single-heterostructure configuration the injected electrons are confined to the p-type region in the vicinity of the n–p interface, increasing the radiative efficiency through stimulated emission. In the double-heterostructure geometry the injected holes and electrons are both confined, increasing their density, increasing the stimulated emission, and increasing external light yield through confinement by reflection from the $Al_xGa_{1-x}As$–GaAs interfaces.

diffusion away from the active region. Thus by adjusting the doping levels in the central slice of GaAs and the adjacent alloy sheets, the larger energy gap in the latter can be used as a barrier against attrition of the injected carriers. The increase in recombination efficiency is measured by the thickness of the central slice (of order 0.5 μm) compared to the distance of several microns that the injected carriers might diffuse were it not for the alloy barriers.

There is a bonus connected with the presence of the alloy barriers. The light emitted in the GaAs slice has a better chance of escaping from the crystal without being absorbed because the energy gap in the alloy barriers is larger than that in the GaAs slice. Moreover, because of the relation between the index of refraction $n = [\epsilon_1(0)]^{1/2}$ and the optical energy gaps [see Eq. (7.3)], the larger energy gaps in the alloy mean a smaller index of refraction. This gives rise to some internal reflection of light generated in the central slice at the interfaces with the alloy barriers, which increases the intensity of the light in the central region and reduces the threshold current for lasing action. In this way it is possible to fabricate injection lasers which operate continuously at room temperature with a 1.5 to 2 V bias.

INTERVALLEY TRANSFER OSCILLATORS

The IMPATT diodes are examples of microwave oscillators based on negative differential resistance regions of the I vs. V characteristic of inhomogeneous semiconductors. It is also possible to achieve negative differential resistance (NDR) in homogeneous samples with hot carriers produced by an electric field. The simplest and most important mechanism for achieving NDR was proposed by Ridley and Watkins in 1961, and realized by Gunn in 1963.

The Ridley–Watkins mechanism assumes that the semiconductor is, for example, n-type and that there are two conduction-band valleys. The one that is lower in energy has a small effective mass and carriers in that valley have a high mobility. The higher conduction band valley is centered on a different point in **k** space and has a much larger effective mass, and consequently a lower mobility. The bottoms of the two conduction bands are separated in energy by an amount δE somewhat greater than the kinetic energy acquired by electrons in large electric fields below the breakdown limit. An example is GaAs (Fig. 5.13) where at room temperature the valley at Γ_1 is lowest, with effective mass $m_1^* \sim 0.07m$, while the X_1 valley is $\delta E \sim 0.36$ eV higher in energy with $m_2^* \sim 1.2m$.

In the absence of an electric field almost all the electrons will be bound to donors or concentrated in the Γ_1 valley. A few will be in the X_1 valley, but the Boltzmann factor $\exp(-\delta E/kT)$ is so small at room temperature ($kT \sim 0.025$ eV) that the number in the X_1 valley will be very small. When a field is applied, population redistribution occurs as the electrons acquire kinetic energy and are scattered from the low-mass valley Γ_1 to the high-mass valley X_1. Increasing the field further populates the X_1 valleys and reduces carrier mobility. If competing processes (such as impact ionization or carrier injection) do not supply more carriers, the current may decrease and hence one has a region of NDR. Even if the sample does not exhibit NDR at low frequencies, it may still do so at microwave frequencies, because intervalley scattering is more rapid than some of the competing damping processes.

The central limitation of intervalley transfer oscillators is bunching of carriers in space to form domains of high carrier density that short out the applied field. Thus the sample becomes effectively inhomogeneous, as it contains regions of high field (above the NDR portion of the I vs. V characteristic) with few carriers, and regions of low field (below the NDR fields) with large numbers of carriers [6]. The samples may still exhibit oscillations with period determined by the transit time of the space-charge domains, but the oscillations are nonsinusoidal and difficult to stabilize, and much of the NDR is wasted.

Hilsum has shown [7] that InP may be superior to GaAs as a material for intervalley transfer oscillators. The reason is that three valleys are partially occupied at high fields in InP, and the transfer of carriers between valleys is such as to suppress spatial regions of high electric field gradient. This inhibits domain formation and makes possible higher frequencies and greater stability.

SEMICONDUCTORS AND MATERIALS SCIENCE

Everyone agrees that practical applications of our understanding of nature may be of great value to society, and in no field are such applications more promising than in materials science. In practice, however, there are many barriers to translation of fundamental knowledge into practical, working devices. Often such developments require new technology, and the number of people with the ability to implement ideas that are technologically feasible is very much smaller than the number needed.

At present (1973) semiconductor technology is very much focused on integrated circuits fabricated on silicon chips. Thus one may hear it said that "if you can't do it with silicon, it isn't worth doing." This is a colloquial exaggeration, of course, but it may well be descriptive of the technology of the 1970s.

Looking further into the future, one may speculate that the great sophistication of transistor technology will have a growing impact on materials science as a whole, and that more and more use will be made of semiconductor techniques outside the realm of integrated circuits. There are many technological applications of metals and insulators which might utilize degenerate semiconductors or large-gap semiconductors, respectively, and thereby gain greater flexibility as well as greater stability and reliability. For these applications one should keep in view the entire field of semiconductor materials, as we have attempted to do in this book.

REFERENCES

1. S. M. Sze, "Physics of Semiconductor Devices." Wiley (Interscience), New York, 1969.
2. J. Bardeen and W. H. Brattain, *Phys. Rev.* **75,** 1208 (1949).
3. R. W. Keyes, *Comments Solid State Phys.* **IV,** 12 (1971).
4. D. G. Thomas, *Phys. Today*, p. 42 (February, 1968).
5. M. B. Panish and I. Hayashi, *Sci. Amer.* **224,** No. 1, 32 (1971).
6. E. M. Conwell, *Phys. Today*, p. 35 (June 1970).
7. C. Hilsum and H. D. Rees, *Proc. Int. Conf. Semiconduct. 1970.* Nat. Tech. Inform. Serv., Nat. Bur. of Std. CONF-700801.

Author Index

Subject Index

A

Acoustic lattice modes, 81
  longitudinal (la), 82
  transverse (ta), 82
Alloys
  energy bands of, 214
  junction lasing from, 275
  nearly ideal, 205
  pseudobinary, 208, 275
Ambipolar doping, 245
Antisymmetric potential, 31, 136
Atomic orbitals, 113
Avalanche diodes, 266

B

Band structures
  diamond, 106, 116

Ge and GaAs, 116
  gray Sn and HgX, 119
  InSb and InAs, 118
  PbS, 120
  silicon, 116
  simplified, 103
Bands, energy
  bending of, 251
  definition, 2
  degeneracies of, 230
  electromechanical analogies, 4
  folding back of, 101
  language of, 98
  pressure dependence of, 149
Barrier heights, 254
Base depths, 270
Bloch functions, 98
Bond charges, 91–92, 142

284